Introduction to PROOFS in MATHEMATICS

Introduction to PROOFS in MATHEMATICS

JAMES FRANKLIN
University of New South Wales

ALBERT DAOUD
University of New South Wales

PRENTICE HALL

New York London Toronto Sydney Tokyo

Typeset by Monoset Typesetters, Strathpine, Queensland.

Printed and bound in Australia by by Brown Prior Anderson Pty Ltd, Burwood, Vic.

4 5 92

ISBN 0-13-474313-X

National Library of Australia
Cataloguing-in-Publication Data

Franklin, J. W. (James William), 1953-
Introduction to proofs in mathematics.

Includes index.
ISBN 0-13-474313-X

1. Proof theory. I. Daoud, Albert Tatar, 1936-
II. Title.

511.3

Library of Congress
Cataloguing-in-Publication Data

Franklin, James, 1953-
Introduction to proofs in mathematics.
Sharam Hekmatpour.

1. Proof theory. I. Daoud, Albert, 1936-
II. Title.

QA9.54.F68 1989 89-25551

ISBN 0-13-474313-X

Prentice Hall, Inc., *Englewood Cliffs, New Jersey*
Prentice Hall Canada, Inc., *Toronto*
Prentice Hall Hispanoamericana, SA, *Mexico*
Prentice Hall of India Private Ltd, *New Delhi*
Prentice Hall International, Inc., *London*
Prentice Hall of Japan, Inc., *Tokyo*
Prentice Hall of Southeast Asia Pty Ltd, *Singapore*
Editora Prentice Hall do Brasil Ltda, *Rio de Janeiro*

PRENTICE HALL

A division of Simon & Schuster

Contents

Preface vii

1 Proof 1
2 "All" statements 12
3 "If and only if" statements 25
4 "Some" statements 32
5 Multiple quantifiers 39
6 "Not", contradiction and counterexample 48
7 Sets 61
8 Proof by mathematical induction 76
9 Vector spaces 86
10 Limits and convergence 100
11 Counting 116
12 Functions 123
13 Real-life problems 131
14 Non-deductive logic 140
15 Axioms, symbolic logic and insight 143

Appendix Application to computer programs 150

Solutions to selected exercises 154

Preface

University teachers of mathematics agree that their students' lack of knowledge is not uniformly distributed across topics, proofs being, however, usually at the head of the list of problems. Why do students take the instruction, "Prove ... " in examinations to mean, "Go to the next question"?

Students of mathematics and computing need to learn how to understand and construct proofs. Proofs are central to mathematics—and more so as the level of mathematics rises. In computer science, it has become clear with the increased importance of software that a computer program is a logical object, the understanding of which requires the same tools as mathematical proof. Proof, however, has not been well served by textbooks. While students do not have the background their teachers had in school of deductive geometry, nothing has replaced Euclid. Textbooks on mathematics for computer science are plentiful, but, as stated in *Carnegie-Mellon Curriculum for Undergraduate Computer Science* (ed. M. Shaw, Springer-Verlag, New York, 1985) on p. 81, their chief problem is "inadequate treatment of logic".

The present book aims to fill this gap. Some of its features are:

1. Techniques of proof are introduced in the context of mathematics, such as arithmetic, already familiar to the student.
2. The techniques are later applied to the usual topics of first-year university mathematics, such as linear algebra and calculus.
3. At all times, matters of context and motivation are kept in mind.
4. Issues of strategy and tactics in constructing proofs are emphasised.
5. The skills of setting out proofs to result in a convincing argument are insisted upon.
6. The continuity of proof with ordinary mathematical techniques, such as calculation, is respected.
7. There is no discussion of symbolic logic, as there are enough books on this subject already.
8. Since the propositional calculus has little importance in mathematical proof, it is barely mentioned. The quantifiers "all" and "some" are the central concern.

We owe a great debt to Jack Gray who read the manuscript in full and made many very valuable suggestions. Useful suggestions were also made by Rod James, John Loxton and David Hunt, as well as by Daniel Solow, whose book *How to Read and Do Proofs* (Wiley, New York, 1982) has broadly the same approach as this book. We would also like to thank Rose Gonzalvez for her fast and efficient typing of the manuscript.

J. Franklin
A. Daoud

1
Proof

In mathematics things are proved; in other subjects they are not.

This statement needs certain qualifications, but it does express the most obvious difference between mathematics and the other sciences. In most fields of study, knowledge is acquired the hard way—from observations, by reasoning about the results of observations and by studying the observations, methods and theories of others. Mathematics was once like this too. Ancient Egyptian, Babylonian and Chinese mathematics consisted of rules for measuring land, computing taxes, predicting eclipses, solving equations, and so on. Methods were learnt from the observations and handed down to others. Modern school mathematics is still often practised in this way.

There were changes in the approach to mathematics. The ancient Greeks found that in arithmetic and geometry it was possible to *prove* that results were true. They found that some truths in mathematics were obvious and that many of the others could be shown to follow logically from the obvious ones. Pythagoras' theorem on right-angled triangles, for example, is not obvious, but a way was found of deducing it from geometrical facts that were apparent. At first it was hoped that every subject would become like mathematics, with all the truths following from obviously true basic statements. This did not happen. Physics, biology, economics and other sciences discover general truths, but to do so they rely on observations. The theory of relativity is not *proved* true; it is tested against observations.

As a result, mathematics has always been regarded as having a different kind of certainty to that obtainable in other sciences. If a scientific theory is accepted because observations have agreed with it, there is always in principle a small doubt that a new observation will not agree with the theory, even if all previous observations have agreed with that theory. If a result is proved correctly, that cannot happen. For more than two thousand years mathematics has attracted those who valued certainty, and has served as the supreme example of certain knowledge. It has also attracted those who wanted knowledge that did not rely on the authority of others; a moment's thought will reveal how little of our knowledge is like this.

Can we be sure, however, that the steps in our reasoning are correct? Are we really sure that what seems obvious to us is true? Can we expect *all* mathematical truths to follow from the obvious ones? These questions are not easily answered, and must be left until after some examples of proofs have been provided. Let us now look at the construction of proofs. We begin with an easy and short proof.

Example 1

Show that,

$$\frac{1}{1{,}000} - \frac{1}{1{,}001} < \frac{1}{1{,}000{,}000}$$

Proof

$$\frac{1}{1{,}000} - \frac{1}{1{,}001} = \frac{1{,}001}{(1{,}000).(1{,}001)} - \frac{1{,}000}{(1{,}000).(1{,}001)}$$

$$= \frac{1}{(1{,}000).(1{,}001)} = \frac{1}{1{,}001{,}000}$$

but,

$$1{,}001{,}000 > 1{,}000{,}000$$

so,

$$\frac{1}{1{,}001{,}000} < \frac{1}{1{,}000{,}000}$$

therefore,

$$\frac{1}{1{,}000} - \frac{1}{1{,}001} < \frac{1}{1{,}000{,}000}$$

Notes

1. Creating this proof consisted of two main parts: Firstly, after noting that the result was not obviously true, we applied a standard technique to,

   $$\frac{1}{1{,}000} - \frac{1}{1{,}001}$$

   namely the technique for subtracting fractions. The result of this was the number $\frac{1}{1{,}001{,}000}$.

   Secondly, comparing where we were with where we wanted to be, we realised that we had yet to show that,

   $$\frac{1}{1{,}001{,}000} < \frac{1}{1{,}000{,}000}$$

   It is clear that the reason this is true is that the bottom line on the left is greater than the bottom line on the right. By writing this down, we have completed the proof.
2. We could have checked the answer by calculating both sides using a calculator or computer and comparing the answers. Why not do this, instead of going to the trouble of finding a proof? The main reason is that mathematics looks for understanding. A calculation in a machine gives an answer, but no understanding of why that answer is right. A proof can make it clear *why* that answer is true—as it did in the example above. Not only is understanding good in itself, it also has practical advantages. For example, if we next want to know about the size of,

$$\frac{1}{2{,}000} - \frac{1}{2{,}001}$$

the calculation that checked that,

$$\frac{1}{1{,}000} - \frac{1}{1{,}001} < \frac{1}{1{,}000{,}000}$$

would be of no help at all, for we would just have to start calculating again. We can, however, see that a proof similar to the one above would show that,

$$\frac{1}{2{,}000} - \frac{1}{2{,}001} < \frac{1}{(2{,}000)^2}$$

In fact there is a similar proof for any whole number n. It would show that,

$$\frac{1}{n} - \frac{1}{n+1} < \frac{1}{n^2}$$

for *all* whole numbers n. (Proofs of such "all" statements are the subject of the Chapter 2.) Understanding, then, allows us to generalise. When we understand a proof we can examine it to see which aspects of the problem are really relevant to the argument, and which are not. In the above proof, while it was important that the reciprocals of two consecutive numbers were subtracted, it was of less importance what exactly these numbers were.

Computers have further limitations. There is a limit to the size of numbers they can work with, whereas a proof can deal with numbers of any size. Furthermore, although beginners in computing believe that everything the computer calculates is correct, this is not true. Because of such problems as the accumulation of round-off errors, it is in the nature of computers to produce wrong answers some of the time. "Garbage in, garbage out", as the experts say, but also occasionally, "Perfect figures in, garbage out". A simple example can be found in A. W. Roberts, *Elementary Linear Algebra* (2nd edn) (1985), Benjamin/Cummings, Reading, Mass., p. 15. The art of telling how much of computer output is true requires exactly the same kind of thinking as is required for proofs (see Appendix for some examples).

3. The proof derived a result that was not obvious, namely,

$$\frac{1}{1{,}000} - \frac{1}{1{,}001} < \frac{1}{1{,}000{,}000}$$

using a sequence of steps, each of which was obviously correct. However, whatever follows logically from what is obviously true, must itself be true. In this case there is no difficulty in knowing that the steps are obvious, since we have used only easy arithmetical operations at each stage. In general, however, what counts as "obvious" and "obviously follows" is something the student has to learn from experience—though in fact most people intuitively agree about what is obvious. Still, even in advanced mathematics, there are occasionally disputes about what is obvious. There are many anecdotes of mathematics professors stopping in the middle of lectures after saying, "This step is obvious.", thinking for 15 minutes, then continuing with "Yes, it is obvious."

Disputes and mistakes about what is obvious could, in principle, be avoided by laying down for each branch of mathematics certain basic statements, called *axioms*, and agreeing that proofs must be derived from these. For example, we could insist that statements of arithmetic must be proved from the basic number laws (such as the commutative law of addition: $a + b = b + a$ for all real numbers a and b). (Axioms are discussed further in Chapter 15.) In practice, axioms are usually not referred to explicitly. Instead, a wide variety of well-known facts and techniques, such as those used in the proof above, are taken to be obvious.

The next proof is an easy one to understand, but a difficult one to find.

Example 2

Show that the sum of the first hundred whole numbers is 5050.

Proof

We have to show that,

$$1 + 2 + 3 + 4 + 5 + \ldots + 99 + 100 = 5050$$

Instead of adding the numbers from left to right, we add the first and the last, then the second and the second to last, and so on.

So,

$$1 + 2 + 3 + 4 + \ldots + 99 + 100 = (1 + 100) + (2 + 99) + (3 + 98) + \ldots + (50 + 51)$$

Now each pair in brackets adds up to 101. So there are fifty 101s. So the sum of the first hundred numbers is,

$$50 \times 101 = 5050$$

Notes

1. Again, an answer could have been obtained on a calculator, but only after much effort. The point is, however, to find a method that makes the answer clear, and brute calculation cannot do this.
2. Adding a hundred randomly chosen numbers *would* require a calculator, but here we have the opportunity to use the fact that the numbers given exhibit a *pattern*: they are the *first* hundred numbers, with each number being one more than the one before it. The essential idea of the proof was to take advantage of this pattern by pairing the first and the last, the second and the second to last, etc. (Some other ideas are given in Exercise 1.13 below.) This step is definitely a hard one to create, with no rules that can be followed in order to find such steps. Some intellectual creativity is needed. In a sense, finding proofs is like playing chess; there are some rules, such as the number laws, and some commonly-used tactics, such as adding numbers in a different order. Overall strategy is a combination of creativity and techniques learnt from experience.
3. The following method used to derive the result $1 + 2 + 3 + \ldots + 100 = 5050$, does not count as a proof:

   "The formula for the sum of an arithmetic progression,

   $$a + (a + d) + (a + 2d) + (a + 3d) + \ldots + (a + (n - 1)d)$$

   is,

   $$na + \frac{n(n - 1)}{2}d$$

   Putting,

   $$a = 1,\ d = 1,\ n = 100$$

   we get,

   $$1 + 2 + 3 + \ldots + 100 = 100.1 + \frac{100(99)}{2}.1 = 5050"$$

The reason this is not a full proof is that it does not deduce the result to be proved from something obvious; the formula for the sum of an arithmetic progression is, if anything, less obvious than the result to be proved. This defect could be repaired by adding a proof of the formula, but in fact this proof is just a more general version of the proof given for $1 + 2 + 3 + \ldots + 100$.

4. It is again natural to look for generality. It is clear that there is no reason why the numbers to be added should stop at 100; it would, therefore, be natural to use the method of proof to derive a formula for $1 + 2 + 3 + \ldots + n$, for *any* whole number n. (See Chapter 2 for further developments.) A further analysis of the proof will also show that the numbers do not need to start at 1, and the gap between one number and the next need not be 1. This will lead to the general formula for the sum of an arithmetic progression.

In the next example, although it takes some effort to find the proof, it is possible to explain and learn the methods for finding it. It does not need a major "bright idea" as in the last example.

Example 3

Prove that, $\sqrt[8]{8!} < \sqrt[9]{9!}$ (where, for any whole number n, n!, read, "n factorial", is the product of the whole numbers up to n, that is, $1.2.3. \ldots .n$).

Finding the proof

The general strategy should be to do something to both sides that will make the problem simpler. A much easier example of a similar kind would be:

Show,

$$\sqrt{2} < \sqrt{3}$$

The reason this is true is that $2 < 3$; that is, one compares the *square* of both sides because $\sqrt{2}$ *means*, "the number whose square is two", so that to be able to deal with it, it is most natural to consider its square. The complete proof of $\sqrt{2} < \sqrt{3}$ would then start with the obviously true fact:

$$2 < 3$$

and, by taking the square root of both sides, arrive at the result to be proved:

$$\sqrt{2} < \sqrt{3}$$

The case here is similar. Because $\sqrt[8]{8!}$ means the (positive) number whose 8th power is 8!, it is natural to consider the 8th power of both sides. This would give,

$$(\sqrt[8]{8!})^8 < (\sqrt[9]{9!})^8$$

that is,

$$8! < (\sqrt[9]{9!})^8$$

If we could show this we could take the 8th root of both sides to show the original result. Unfortunately, while this step has made the left-hand side simpler, it has made the right-hand side more complicated. Similarly, if we took the 9th power of both sides to remove the $\sqrt[9]{\ }$, we would get,

$$(\sqrt[8]{8!})^9 < 9!$$

which is an advance on the right, but not so on the left. To overcome this we look at a power of both sides which will cause the roots to disappear on both sides. We take the 72nd power, since $8 \times 9 = 72$. That is, we will try to show,

$$(\sqrt[8]{8!})^{72} < (\sqrt[9]{9!})^{72}$$

which is, on simplifying,

$$(8!)^9 < (9!)^8$$

This is still not obviously true, but using the meanings of 8th and 9th powers (namely, x^8 means x.x. . . . eight times, and similarly x^9 means x.x. . . . nine times), we are trying to show,

$$\underbrace{8! \ldots 8!}_{\text{nine times}} < \underbrace{9! \ldots 9!}_{\text{eight times}}$$

As this is still not obviously true, we apply the definitions of 8! and 9!. We wish to show,

$$\underbrace{(1.2. \ldots .8)(1.2. \ldots .8) \ldots (1.2. \ldots .8)}_{\text{nine times}} < \underbrace{(1.2. \ldots .9)(1.2. \ldots .9) \ldots (1.2. \ldots .9)}_{\text{eight times}}$$

It is clear that most of the numbers appear on both sides. Each factor (1.2. . . . .9) on the right can be thought of as (1.2. . . . .8).9, so (1.2. . . . .8) appears nine times on the left and eight times on the right. Dividing both sides by $(1.2. \ldots .8)^8$ gives,

$$1.2. \ldots .8 < \underbrace{9. \ldots .9}_{\text{eight times}}$$

If we could show this, we could multiply it by $(1.2. \ldots .8)^8$ to get the result above. But now we have reached a result which *is* obviously true, since both sides have eight factors, and the ones on the left are all less than the ones on the right.

The final proof is obtained by writing these steps in *reverse* order, since the aim is to derive the result to be proved *from* something obvious, not vice versa (see Note 2 below).

Proof

$1.2. \ldots .8 < \underbrace{9.9. \ldots .9}_{\text{eight times}}$ since 1, 2, . . . , 8 are all less than 9.

Multiplying both sides by $(1.2. \ldots .8)^8$,

$$\begin{aligned}(1..2. \ldots .8)^9 &< (1.2. \ldots .8)^8.9^8 \\ &= (1.2. \ldots .9)^8\end{aligned}$$

that is,

$$(8!)^9 < (9!)^8$$

Taking the (positive) 72nd root of both sides,

$$\sqrt[8]{8!} < \sqrt[9]{9!}$$

Notes

1. The proof as now written is designed to convince the reader of the truth of the conclusion. However, it gives no understanding of how the proof was derived. No explanation is given, for example, of why the first line should be what it is. This is the usual style in written mathematics. The advantage of such a style is that it communicates results and produces confidence in them in the shortest possible space; its disadvantage is that it covers up a certain amount of what is going on, by writing down assertions for which the motivation is totally unclear.
2. The reason for writing the steps in reverse order is that we must derive the result to be proved from something obvious, not derive something obvious from the result to be proved. The point is clear from the following fallacious argument:

   Show that $1 = 3$

   "Proof": Suppose,

   $$1 = 3$$

   so,

   $$-1 = 1$$

   (subtracting 2 from both sides).

   So,

   $$1 = 1$$

   (squaring both sides).

   This is true, so,

   $$1 = 3$$

   The fallacy of this "proof" is that it has derived correctly something obviously true (namely $1 = 1$) from the result to be proved ($1 = 3$) but has then wrongly concluded that this means the result to be proved is also true. A correct proof must *end* with the result to be proved.
3. The need to reverse steps, in some cases, means that care is needed in checking that the steps involved are in fact reversible. For example, the step,

   $$\sqrt{2} < \sqrt{3}$$

   therefore,

   $$\sqrt{2} + 2 < \sqrt{3} + 2$$

   is reversible, in that one can correctly argue,

   $$\sqrt{2} + 2 < \sqrt{3} + 2$$

   therefore,

   $$\sqrt{2} < \sqrt{3}$$

   (because 2 has been subtracted from both sides).

   Most simple arithmetical steps are reversible like this, but some are not. For example, one can correctly argue,

   $$x = 2$$

   therefore,

   $$x^2 = 4$$

   However, one cannot reverse this step to argue,

   $$x^2 = 4$$

therefore,

$$x = 2$$

Rather, it would be only correct to argue,

$$x^2 = 4$$

therefore,

$$x = 2 \quad \text{or} \quad x = -2.$$

Otherwise, fallacious arguments like the following would result:

$$\text{“}(-1)^2 = 1$$

but,

$$1 = 1^2$$

so,

$$(-1)^2 = 1^2$$

Taking the square root of both sides,

$$-1 = 1\text{”}$$

4 Could the search for a proof have started in some other way? For example, could we have started with:

$$\begin{aligned}\sqrt[8]{8!} &= \sqrt[8]{1.2.\ \ldots\ .8}\\ &= \sqrt[8]{1}.\sqrt[8]{2}.\ \ldots\ .\sqrt[8]{8}\end{aligned}$$

In fact a proof can be found in this way, with the resulting proof not being very different to the one found above. The reader may like to try it. As in chess, one can learn by experience which moves tend to be good; in general, however, there is only a limited amount of advice which can be given. Some general guidelines which could be given are: “Do something that makes things simpler.”; and “Look for some step that will make the current situation more like the goal.” A technique that was used a number of times in the last example was what could be roughly described as, “expand the definitions”. To deal with $\sqrt[8]{x}$, we used the definition of an 8th root, namely,

$$\sqrt[8]{x} \text{ is the number } y \text{ such that } y^8 = x$$

to deal with x^9, we replaced it with,

$$\underbrace{x\ \ldots\ x}_{\text{nine times}}$$

which is what x^9 means; and to deal with n!, we replaced it by,

$$1.2.\ \ldots\ .n$$

In each of the examples above a true result was given and a proof was asked for. So, where did the result itself come from? Which came first, the proof or the result? How would anyone know that 1/1,000 − 1/1,001 was less than 1/1,000,000 unless they had already performed the calculation given in the proof above? Realistically, the proof and the result are often discovered together. If the following two questions are asked, “About how big is a number like 1/1,000 and 1/1,001. Wouldn't it be much smaller than the numbers themselves?”, the natural thing would be to perform the calculation of 1/1,000 − 1/1,001 and find that the answer was smaller than 1/1,000,000. The result and the proof would be found simultaneously, and the discovery would then be presented as the result, followed by its proof. On the other hand, it often happens that a result is found or conjectured without a proof, with a subsequently found proof giving an understanding of why the result is true. For example, someone might add up $1 + 2 + 3 + \ldots + 100$ with a calculator, find the answer 5050, notice this is 50×101, and ask why this should be so. The above proof provides the answer. (The matter of conjectures that are not yet proved will be discussed further in Chapter 14.)

Exercises

(Gradings of exercises: * easy, ** moderate, *** difficult.)

*1. Show that: $\frac{1}{1{,}000} - \frac{1}{1{,}002} < \frac{2}{1{,}000{,}000}$

*2. Show that: $1 + 2 + 3 + \ldots + 999 + 1{,}000 = 500{,}500$

*3. Show that: $\sqrt{1{,}001} - \sqrt{1{,}000} < \frac{1}{2\sqrt{1{,}000}}$

Hint: Multiply $\sqrt{1{,}001} - \sqrt{1{,}000}$ by $\frac{\sqrt{1{,}001} + \sqrt{1{,}000}}{\sqrt{1{,}001} + \sqrt{1{,}000}}$

In questions 4, 5, 8, 14 and 21 it will be necessary to assume common trig. formulas such as, $\sin(A + B) = \sin A \cos B + \cos A \sin B$. These are not obvious, but are at least well-known.

*4. Given that: $\sin A = \frac{3}{5}$ and $\sin B = \frac{5}{13}$ and that A is obtuse (i.e. between 90° and 180°) and B is acute, show that: $\cos(A + B) = -\frac{63}{65}$

*5. Show that: $\tan\frac{\pi}{12} = 2 - \sqrt{3}$

*6. Prove that: $x = 1 - \sqrt{5}$ is a solution of the equation $x^3 - 3x^2 - 2x + 4 = 0$

*7. (a) Show that: $\log_5 1^2 + \log_5 2^2 + \log_5 3^2 - 2\log_5 6 = 0$

(b) Show that: $\log_4((\sqrt{2})^9) = \frac{9}{4}$

*8. Show that: $\cos\frac{5\pi}{12} = \frac{\sqrt{3} - 1}{2\sqrt{2}}$

*9. Prove that: $\sqrt[7]{7!} > \sqrt[6]{6!}$

*10. Comment on the following reasoning: "If a pond is 1 metre across, how far is it around the edge? Answer: circumference $= \pi d$ metres $= 3.14159265$ metres (correct to 8 decimal places)."

****11.** The Golden Ratio is defined to be the number ℓ so that if a 1×1 square is removed from an $\ell \times 1$ rectangle the remaining rectangle is the same shape as the original. (See Figure 1.1.) (The ancient Greeks considered rectangles of this shape especially well proportioned, and used this shape in architecture.)

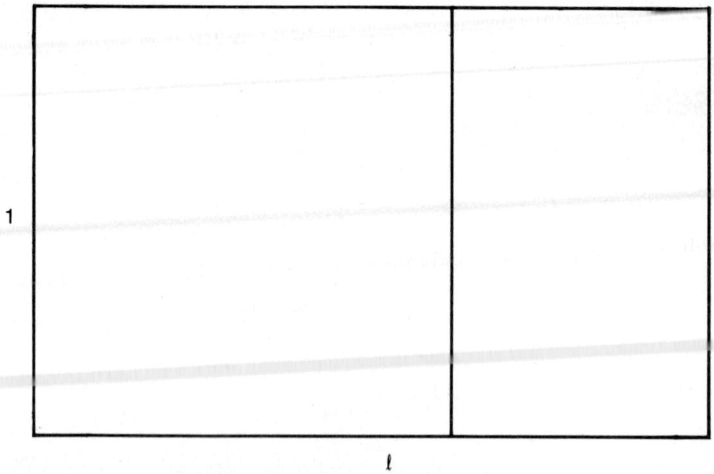

Figure 1.1

Prove that: $\ell^2 = \ell + 1$ and hence find an expression for ℓ.

****12.** Show that: $\sqrt{2 + \sqrt{2}} + \sqrt{2 - \sqrt{2}} < 2\sqrt{2}$

****13.** Discuss whether the following proof of: $1 + 2 + 3 + \ldots + 99 + 100 = 5{,}050$ is better or worse than the one given in Example 2.
It might be useful to see how the method works with $1 + 2 + 3 + \ldots + 99$:

$$2(1 + 2 + 3 + \ldots + 99 + 100) = \begin{array}{r} 1 + 2 + 3 + \ldots + 99 + 100 \\ + 100 + 99 + 98 + \ldots + 2 + 1 \end{array}$$

(since in the second row we have just written down the numbers in reverse order)

$$= \underbrace{101 + 101 + \ldots + 101 + 101}_{100 \text{ times}}$$

(since each number in the top row plus the one below it gives 101)

$$= 100 \times 101$$

so,

$$1 + 2 + 3 + \ldots + 99 + 100 = \frac{100 \times 101}{2} = 5{,}050$$

****14.** (a) Prove that: $\cos 3\theta = 4\cos^3\theta - 3\cos\theta$

(b) Hence show that $\cos\frac{2\pi}{3}$ is a root of the equation $4x^3 - 3x - 1 = 0$

**15. Prove that: π is between 3 and 4 (where π is defined to be the ratio of the circumference of a circle to its diameter; that is, π is defined by:

$$\text{Circumference} = \pi \times \text{diameter})$$

Hint: See Figure 1.2.

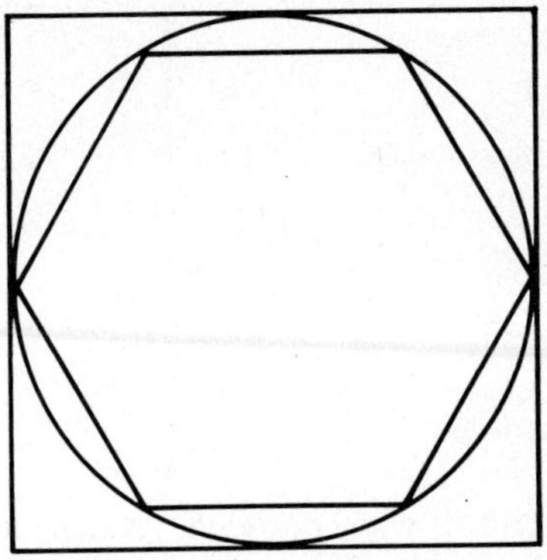

Figure 1.2

**16. Prove that the second digit after the decimal point of $\sqrt{2}$ is 1. (Calculators not allowed.)

**17. Show that: $\sqrt[3]{3 + \sqrt[3]{3}} + \sqrt[3]{3 - \sqrt[3]{3}} < 2\sqrt[3]{3}$

***18. Prove that $0 < \left(1 + \frac{1}{1000}\right)^{1000} < 3$

Hint:

$$\left(1 + \frac{1}{n}\right)^n = 1 + \frac{n}{n} + \frac{n(n-1)}{1.2}\cdot\frac{1}{n^2} + \frac{n(n-1)(n-2)}{1.2.3}\cdot\frac{1}{n^3} + \ldots + \frac{n!}{n!}\left(\frac{1}{n}\right)^n$$

$$< 1 + 1 + \frac{n.n}{1.2}\cdot\frac{1}{n^2} + \frac{n.n.n}{1.2.3}\frac{1}{n^3} + \ldots + \frac{n^n}{n!}\frac{1}{n^n}$$

$$< 1 + 1 + \frac{1}{2!} + \frac{1}{3!} + \ldots + \frac{1}{n!}$$

$$< 1 + 1 + \frac{1}{2} + \frac{1}{2^2} + \ldots + \frac{1}{2^{n-1}}$$

***19. Prove that: $10^9 < 9^{10}$

***20. Prove that: $99^{100} > 100^{99}$

***21. Show that: $\frac{\pi}{4} = \tan^{-1}\left(\frac{1}{4}\right) + \tan^{-1}\left(\frac{3}{5}\right)$

2

"All" statements

The results proved in Chapter 1 were *particular* statements. A particular statement either asserts that a property, or *predicate*, is true of a *subject* (e.g. "seven is prime"; "Socrates is mortal"), or asserts some *relation* between two or more things (e.g., $\sqrt[8]{8!} < \sqrt[9]{9!}$; $1/1{,}000 - 1/1{,}001 < 1/1{,}000{,}000$). In any science, but especially in mathematics, we are interested in *generalising* from particular facts. When we saw that,

$$\frac{1}{1{,}000} - \frac{1}{1{,}001} < \frac{1}{1{,}000{,}000}$$

we realised that something similar would have happened whenever we subtracted the reciprocals of two consecutive whole numbers. A similar proof would have shown, for example, that,

$$\frac{1}{203} - \frac{1}{204} < \frac{1}{(203)^2}$$

The *particular* result that,

$$\frac{1}{1{,}000} - \frac{1}{1{,}001} < \frac{1}{1{,}000{,}000}$$

is an instance of the *general* pattern: for all whole numbers n,

$$\frac{1}{n} - \frac{1}{n+1} < \frac{1}{n^2}$$

This is an example of an "all" statement, or a "universal generalisation" in the jargon of logic. (Unfortunately, logic is a subject especially given to producing jargon; a precise technical language is sometimes needed, but should not be overdone.)

Some equivalent ways to express the same result in English are:

For every whole number n,

$$\frac{1}{n} - \frac{1}{n+1} < \frac{1}{n^2}$$

For any whole number n,

$$\frac{1}{n} - \frac{1}{n+1} < \frac{1}{n^2}$$

If n is a whole number, then,

$$\frac{1}{n} - \frac{1}{n+1} < \frac{1}{n^2}$$

The difference of the reciprocals of any two consecutive whole numbers is less than the reciprocal of the square of the smaller number.

Some other examples of "all" statements are:

- All men are mortal. (This example has been traditional since the time of the ancient Greeks.)
- All multiples of ten are multiples of five.
- For all whole numbers n,

$$\sqrt[n]{n!} < \sqrt[n+1]{(n+1)!}$$

  (this is the natural generalisation of the particular result in Chapter 1, Example 3: $\sqrt[8]{8!} < \sqrt[9]{9!}$).
- Every fourth power is a square (i.e. if a whole number is the fourth power of some other whole number, it is the square of some whole number). For example, 81 is a fourth power because it is 3^4 and is also a square because it is 9^2.

The general form of an "all" statement is,

All As are Bs

This states that anything that has the property of being an A also has the property of being a B. While some of the above examples cannot be put easily or naturally into the form, "All As are Bs". ("All whole numbers n are such that, $1/n - 1/(n+1) < 1/n^2$", for example, is a little awkward.) The form, "All As are Bs", is nevertheless very useful for explaining the general features of "all" statements.

To prove an "All As are Bs" statement, we must show that *anything* that is an A is also a B. So the proof of, "All As are Bs", should look like this:

Let x be an A

therefore, x is a B.

Example 1

Prove that for all whole numbers n,

$$\frac{1}{n} - \frac{1}{n+1} < \frac{1}{n^2}$$

Proof

Let n be a whole number. Then,

$$\frac{1}{n} - \frac{1}{n+1} = \frac{n+1-n}{n(n+1)}$$
$$= \frac{1}{n^2+n}$$

but,

$$n^2 + n > n^2$$

so,

$$\frac{1}{n^2+n} < \frac{1}{n^2}$$

therefore,

$$\frac{1}{n} - \frac{1}{n+1} < \frac{1}{n^2}$$

Notes

1. Although the result,

$$\frac{1}{n} - \frac{1}{n+1} < \frac{1}{n^2}$$

is much more general than,

$$\frac{1}{1{,}000} - \frac{1}{1{,}001} < \frac{1}{1{,}000{,}000}$$

its proof is no harder. If anything, it is easier, since there is no danger of being blinded by any irrelevant facts about the particular numbers chosen (e.g. their large size, which might easily lead to mistakes in comparing them).

2. It is most important to understand that, to prove an, "All As are Bs" statement, it is *not* sufficient to look at a number of particular As and check that they are Bs. There is no "proof by example". Thus, it would be incorrect to argue,

$$\frac{1}{1} - \frac{1}{2} < \frac{1}{1^2}$$

and,

$$\frac{1}{2} - \frac{1}{3} < \frac{1}{2^2}$$

and,

$$\frac{1}{3} - \frac{1}{4} < \frac{1}{3^2}$$

so,

$$\frac{1}{n} - \frac{1}{n+1} < \frac{1}{n^2}$$

for all n. Checking some examples *supports* the "all" statement, but does *not prove* it; the statement could be true for the examples checked, but false for some others. (This joke told by mathematicians about physicists illustrates the point. A physicist argued that he could show that all odd numbers are prime: "Three and five are prime," he said, "and so is seven. Nine—well, experimental error—but eleven and thirteen are prime, so all odd numbers are prime." Underlying this joke is a real difference between mathematics and experimental sciences such as physics: in physics the "all" statements can only ultimately be confirmed by a sufficiently large number of experiments, while in mathematics they can be *proved* true.)

Example 2

Show that every whole number that is a fourth power is a square.

Proof

We should begin the proof by taking a number that is a fourth power, and writing in symbols what this means:

Let x be a whole number that is a fourth power. That is, $x = a^4$ for some number a.

We wish to conclude that x is in fact a square, so we ask how we can write a^4 as the square of something. As a^4 is a.a.a.a, it can be written $(a.a)^2$, that is $(a^2)^2$. So the complete proof is:

Let x be a whole number which is a fourth power. So,

$$x = a^4$$

for some whole number a

$$= a.a.a.a$$
$$= (a^2)^2$$

So x is a square.

Every fourth power is therefore a square.

Notes

1. As already pointed out, it is not sufficient to check that the result is true for some particular xs that are fourth powers, for example, 16 and 81. The proof must be general, that is, it must apply to *any* x that is a fourth power.
2. The step after, "x is a whole number which is a fourth power" expresses in symbols what it means to be a fourth power. It often happens that in a proof of, "All As are Bs", the second line (after, "Let x be an A") expresses the defining characteristic of As, that is, what it means to be an A. Similarly, the second last line (before, "therefore, x is a B") is often an expression in symbols which shows that x *is* a B. Check this in the example above.
3. The sentence, "Let x be an A", is sometimes expressed as, "Take a general (or arbitrary) A". This expression is possibly misleading, since all actual As are particular. It can be useful, however, as a reminder that in the proof of "All As are Bs", we must use only facts that are true of *all* As.

Watch for these points in the following example.

Example 3

Prove that the square of an odd number is odd. (Obviously, "of an odd number" means "of any odd number".)

Finding the proof

We are asked to prove something about *all* odd numbers, namely, that their squares are odd. So the proof must look like:

Let x be an odd number.

Therefore, x^2 is odd.

To complete the proof it is necessary to express "is odd" in symbols. A number is odd if it is of the form $2k + 1$ for some k; this is the general form of an odd number. So the proof should look like this:

Let x be an odd number.

So, $x = 2k + 1$ for some number k,

so, $x^2 = 2\,(\text{something}) + 1$

therefore, x^2 is odd.

To fill in the steps still missing, we ask how to get from $x = 2k + 1$ to something about x^2. Clearly we should square both sides. The complete proof is as follows:

Proof

Let x be an odd number.

So,

$$x = 2k + 1$$

for some number k,

$$\begin{aligned} x^2 &= (2k + 1)^2 \\ &= 4k^2 + 4k + 1 \\ &= 2(2k^2 + 2k) + 1 \end{aligned}$$

therefore, x^2 is odd.

Note

As occasionally happens, this proof has actually proved a stronger result than was intended. We intended to show that the square of any odd number is of the form $2K + 1$; we actually showed it was of the form $4K + 1$. This is a stronger result, in the sense that the statement

"The square of any odd number is of the form $4K + 1$"

implies the statement,

"The square of any odd number is odd"

but the reverse implication is not true, since some odd numbers (e.g. 999) are not of the form 4K + 1. The stronger result tells us that a number such as 999 cannot be the square of an odd number, while the weaker result does not tell us this.

"All As are Bs" statements are so common that many equivalent ways of expressing them have appeared. We have mentioned already the forms,

"Any A is a B"
"Every A is a B"
"If anything is an A then it is a B"

Some others are:

"Whatever is an A is a B"
"Being an A is a sufficient condition for being a B"
"Something is an A only if it is a B"

(Note that the order of A and B is the *same*: "If A then B"; "A only if B". Think of, "If anything is a horse then it is an animal"; "Something is a horse only if it is an animal".)

"Being a B is a necessary condition for being an A"

(A and B are in the opposite order here. Think of, "Being an animal is a necessary condition for being a horse"—to be a horse, something must at least be an animal.)

"The set of As is included in the set of Bs"

(See Figure 2.1.)

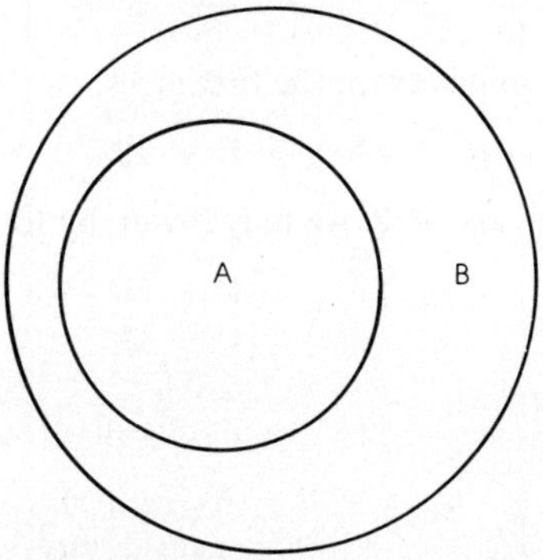

Figure 2.1

Sets will be discussed more fully in Chapter 7. Some ways of expressing "All As are Bs", using "not", will be given in Chapter 6. These statements are used in different situations in English, because they emphasise different things. But while their rhetorical force is different, their logical force is the same: "All As are Bs".

It is most important to understand that "All As are Bs" is not the same as, "All Bs are As". "All horses are animals" is true but, "All animals are horses" is false. The statement "All Bs are As" is called the *converse* of "All As are Bs", and, "If anything is a B then it is an A" is the converse of "If anything is an A then it is a B". If an "all" statement is true, its converse may or may not be true; more work is needed to decide.

Note that there can be "if" statements which are not of the form, "If an x is an A then x is a B". Any two statements p and q can be joined by "if . . . then" to form the statement "If p then q". For example, "If the squares of all odd numbers are of the form 4k + 1, then the squares of all odd numbers are odd". An "if p then q" statement is proved by showing how q follows from p, so its proof is of the form:

In practice, however, almost all the "if . . . then" statements used in mathematics can be thought of in the form "if x is an A then x is a B".

We conclude this chapter with some examples from the theory of linear equations, linear algebra and calculus. They illustrate how important "all" statements are in every branch of mathematics.

Example 4 (Linear equations)

Show that every system of linear equations of the form,

$$\begin{aligned} x + Ay &= B \\ 2x + Cy &= D \end{aligned}$$

such that $C \neq 2A$ can be solved.

Proof

Taking the second equation minus twice the first gives,

$$(C - 2A)\,y = D - 2B$$

since $C \neq 2A$ and so $C - 2A \neq 0$ we may divide by $C - 2A$:

$$y = \frac{D - 2B}{C - 2A}$$

Then from the first equation,

$$\begin{aligned} x &= B - Ay \\ &= B - \frac{A(D - 2B)}{C - 2A} \end{aligned}$$

So the system has been solved.

Example 5 (Linear algebra)

Show that all linear combinations of the vectors (1, 1, 2) and (2, 3, 5) in $\mathbf{R}^3$ lie on the plane,

$$x + y - z = 0$$

Proof

Any linear combination of (1, 1, 2) and (2, 3, 5) is of the form, a(1, 1, 2) + b(2, 3, 5) for some a, b $\in$ **R**. However, this is (a + 2b, a + 3b, 2a + 5b), and this point satisfies the equation,

$$x + y - z = 0$$

since

$$\begin{aligned}(a + 2b) + (a + 3b) - (2a + 5b) &= a + a - 2a + 2b + 3b - 5b \\ &= 0\end{aligned}$$

So every linear combination of (1, 1, 2) and (2, 3, 5) lies on the plane,

$$x + y - z = 0$$

Example 6 (Calculus)

Show that, for all $x > 0$, $x \geqslant \ln x + 1$

Proof

We use calculus methods to compare the graphs of,

$$y = x$$

and

$$y = \ln x + 1$$

The graph of,

$$y = x$$

is the straight line through (0, 0) with gradient 1. Now for $x = 1$,

$$x = \ln x + 1 = 1$$

so the two graphs meet at (1, 1). Also,

$$\frac{d}{dx}(\ln x + 1) = \frac{1}{x}$$

and for all $x > 1$,

$$\frac{1}{x} < 1$$

(and indeed $1/x \to 0$ as $x \to \infty$).

So for $x > 1$, $y = \ln x + 1$ has a gradient less than $y = x$. Thus for $x \geqslant 1$,

$$x \geqslant \ln x + 1$$

For $0 < x < 1$, $1/x$ (the gradient of $y = \ln x + 1$) is greater than 1 (the gradient of $y = x$). So here again

$$x > \ln x + 1$$

So for all $x > 0$,

$$x \geqslant \ln x + 1$$

(See Figure 2.2, which shows these graphs.)

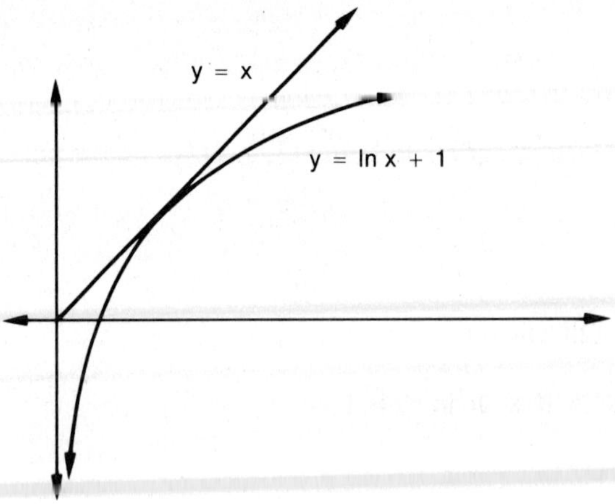

Figure 2.2

Exercises

(Gradings of exercises: * easy, ** moderate, *** difficult.)

*1. Prove that the square of any even number is even.

*2. Prove that the sum of any two consecutive numbers is odd.

*3. Prove that the product of any two odd numbers is an odd number.

*4. Give an "all" statement relating "cows" and "mammals" which illustrates that, "All As are Bs" is not logically equivalent to, "All Bs are As".

*5. Decide whether the statement:

$$x^2 - 3x + 2 < 0$$

for $1 < x < 2$ (i.e. for *all* x between 1 and 2) is true or false and prove your answer.

*6. (a) Rewrite the all statement, "All As are Bs" in the forms
(i) ". . . only if . . ."
(ii) "If . . . then . . ."
(b) A father told his son, "Only if you pass will you get a bike". The son passed but he did not get a bike. Did the father break his promise? Explain.

*7. Show that if m and n are odd integers, then $m + n$ is even.

***8.** Comment on the reasoning:

(a) It takes one person two hours to mow this lawn. So it would take 6,000 people $\frac{2}{6,000}$ hours.

(b) If it takes four ships three days to cross the Tasman, how long will it take seven ships?

***9.** Give one example of an "all" statement (universal generalisation) in the form, "All As are Bs". Then rewrite it in different forms.

***10.** Show that for any whole number n, $\frac{n}{n+1} < \frac{n+1}{n+2}$

***11.** Prove that $\sqrt[n]{n!} < \sqrt[n+1]{(n+1)!}$ for any whole number n (imitate the proof in Chapter 1).

***12.** Prove that the product of three consecutive whole numbers, of which the middle one is odd, is divisible by 24.

***13.** Let α and β be the roots of the quadratic equation $ax^2 + bx + c = 0$, where $a \neq 0$.

Show that:

$$\alpha + \beta = -\frac{b}{a}$$

and,

$$\alpha\beta = \frac{c}{a}$$

(Do not use the quadratic formula.)

****14.** Show that: $\left(1 - \frac{1}{2}\right)\left(1 - \frac{1}{3}\right)\left(1 - \frac{1}{4}\right)\ldots\left(1 - \frac{1}{n}\right) = \frac{1}{n}$

****15.** Prove that if, $0 < x < 1$ then, $0 < x^2 < x < 1$

****16.** Show that for any non-zero real numbers x and y such that $x + y = 1$,

$$\left(1 - \frac{1}{x}\right)\left(1 - \frac{1}{y}\right) = 1$$

****17.** Prove that $x^2 - 4x + 5 > 0$ for all real numbers x.

****18.** (a) Prove that if $a \neq 0$ then;

$$\frac{-b + \sqrt{b^2 - 4ac}}{2a} \quad \text{and} \quad \frac{-b - \sqrt{b^2 - 4ac}}{2a}$$

are solutions of $ax^2 + bx + c = 0$

(b) Prove that if $a \neq 0$ and x is a solution of $ax^2 + bx + c = 0$, then,

$$x = \frac{-b + \sqrt{b^2 - 4ac}}{2a} \quad \text{or} \quad \frac{-b - \sqrt{b^2 - 4ac}}{2a}$$

****19.** Find a generalisation of:

$$\frac{1}{1000} - \frac{1}{1002} < \frac{2}{(1000)^2}$$

and prove it.

****20.** Show that if a, b are integers such that 7 divides $a + b$ and $a^2 + b^2$, then 7 divides both a and b.

****21.** Prove that for all whole numbers n, $(n + 1)(n + 2) \ldots (2n - 1)(2n) = 2^n.1.3.5 \ldots (2n - 1)$

****22.** (a) What is wrong with the following "proof" of Pythagoras' theorem? (The theorem states that if a, b, c are the sides of any right-angled triangle with c the hypotenuse then $a^2 + b^2 = c^2$.)

By the cosine rule,

$$c^2 = a^2 + b^2 - 2ab \cos 90°$$

but,

$$\cos 90° = 0 \quad \text{so} \quad c^2 = a^2 + b^2$$

(b) What about this attempt?
Let θ be the angle opposite b.
Then,

$$\begin{aligned} a^2 + b^2 &= (c \cos \theta)^2 + (c \sin \theta)^2 \\ &= c^2(\cos^2 \theta + \sin^2 \theta) \\ &= c^2 \quad \text{since} \quad \cos^2 \theta + \sin^2 \theta = 1 \end{aligned}$$

****23.** (a) Prove that for any positive real numbers x, y,

$$xy \leqslant \left(\frac{x + y}{2}\right)^2$$

(b) Hence show that of all rectangles with a fixed perimeter, the square has the largest area.

*****24.** (a) Prove that the area, S, of a triangle of base b and altitude h is given by,

$$S = \frac{1}{2}bh$$

Hint: Draw some rectangles (make sure all shapes of triangles are accounted for). See Figure 2.3.

(b) Hence prove that the area S of a triangle is given by,

$$S = \frac{1}{2}ab \sin \theta$$

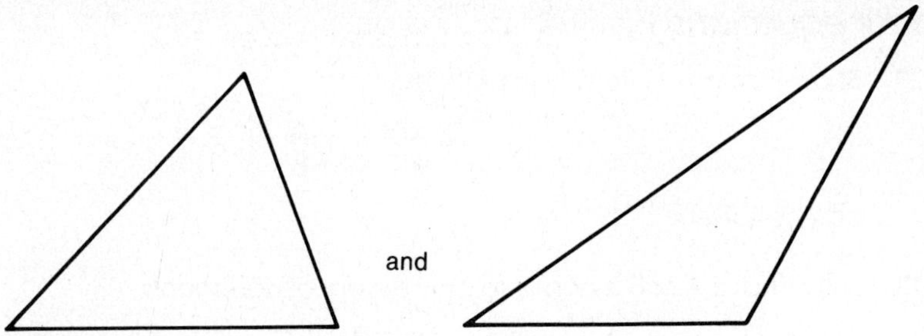

Figure 2.3

where θ is the angle between its two sides a and b.

*****25.** Prove that the area A of a right circular cone is given by,

$$A = \pi r \ell$$

where r is the radius of the base and ℓ is the slant height. (Assume the formula πr^2 for the area of a cirlcle.)
Hint: See Figure 2.4.

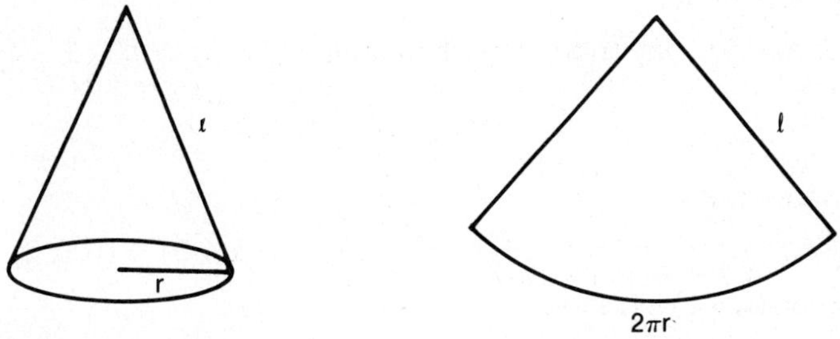

Figure 2.4

*****26.** Prove Pythagoras' theorem: If a, b, c are the sides of any right-angled triangle with c the hypotenuse, then,

$$a^2 + b^2 = c^2$$

*****27.** Show that for any positive integers m and n, 2 is between $\frac{m}{n}$ and $\frac{m + 2n}{m}$ (inclusive).

*****28.** (a) Show that any figure in the XY plane, which is symmetrical about the X axis and symmetrical about the Y axis, is also symmetrical about the origin.
(A figure is said to be symmetrical about an axis if, for any point in the figure, the point opposite it across the axis is also in the figure. Similarly, a figure is symmetrical about a point if, for any point in the figure, the point opposite it across the point of symmetry is also in the figure. Thus a square is symmetrical about both diagonals and also about its centre.)

(b) Find some similar result in three dimensions and prove it.

Linear equations

***29.** Show that any system of equations,

$$x + y = A$$
$$x - y = B$$

can be solved.

****30.** Show that if A or B is non-zero, any system of equations,

$$Ax + By = C$$
$$Bx - Ay = D$$

can be solved.

****31.** Show that any system of equations of the form,

$$x + y + Az = B$$
$$x - y + Cz = D$$

has infinitely many solutions.

Linear algebra

***32.** Show that any linear combination of (1, 2, 3) and (−2, −4, −6) lies on the line,

$$6x = 3y = 2z$$

****33.** Show that if:

$$x(1, 1, 0) + y(1, 2, 3) + z(3, 4, 3) = (0, 0, 0)$$

then (x, y, z) is a scalar multiple of (2, 1, −1)

Calculus

***34.** Prove that for all $x > 0$, $x > \sin x$

****35.** Show that for all $x > 4$, $2^x > x^2$

****36.** Show that any solution of $\frac{dy}{dx} = y$ is a multiple of e^x

****37.** Show that for $x > 10$, the graph of,

$$y = \frac{\sin x}{x^2}$$

lies within 0.01 of the x-axis.

*****38.** Is it true that for all $X > 0$,

$$\int_0^X \frac{\sin x}{x}\,dx > 0?$$

Prove your answer.

3

"If and only if" statements

It was explained in the last chapter that, when an "All As are Bs" statement is true, its converse, "All Bs are As", may or may not be true. If both "All As are Bs" and "All Bs are As" are true, then the As are exactly the same things as the Bs. Such a situation is usually expressed by an "if and only if" statement:

"Something is an A if and only if it is a B"

You will recall, from Chapter 2, that "Something is an A only if it is a B" is equivalent to "All As are Bs", and that "Something is an A if it is a B" is equivalent to "All Bs are As". Therefore, the "if and only if" statement, "Something is an A if and only if it is a B" is equivalent to,

"All As are Bs and all Bs are As"

"If and only if" is sometimes abbreviated to "iff" and symbolised by $\Leftrightarrow$.
In accordance with the use of "necessary" and "sufficient" introduced in Chapter 2, the above "iff" statement may also be expressed as:

"Being an A is a necessary and sufficient condition for being a B"

Some famous "if and only if" statements in mathematics are:

- A triangle has three equal sides if and only if it has three equal angles.
- A polyhedron is regular if and only if it is one of the five Platonic solids (the regular tetrahedron, cube, octahedron, dodecahedron and icosahedron). A polyhedron is said to be *regular* if all its faces are the same, all its edges are the same and all its vertices are the same, that is, at each vertex the same number of edges meet and the angles between them are all the same.
- A number is divisible by 9 if and only if the sum of its digits is divisible by 9.
- A real number is rational if and only if its decimal expansion is terminating or (eventually) repeating.

"If and only if" statements are very highly prized in mathematics. If an, "All As are Bs" statement is found, the mathematician asks, "What conditions are there on a B which will be just enough to make sure it is an A?" If such a condition C is found, the result will be an "if and only if" statement:

"Something is an A if and only if it is a C and a B"

This will completely express the connection between the concepts A and B. An important example is Euler's formula for polyhedra:

$$V - E + F = 2$$

where V is the number of vertices of a polyhedron, E the number of its edges and F the number of faces. This formula is true for most simple polyhedra. It can be checked for the Platonic solids by counting, and, in the eighteenth century, Euler proved that it was true for many others. But finding exactly *which* polyhedra it was true for, that is, finding a condition C such that,

"A polyhedron is a C if and only if $V - E + F = 2$"

proved extremely difficult, and led to many conceptual advances. The story is entertainingly told in Imre Lakatos's book, *Proofs and Refutations* (1976), Cambridge University Press, Cambridge.

Since an "if and only if" statement really makes two assertions, its proof must contain two parts. The proof of "Something is an A if and only if it is a B" will look like this:

Let x be an A,

therefore, x is a B.

Let y be B,

therefore, y is an A.

So, something is an A if and only if it is a B.

Example 1

Prove that a whole number is even if and only if its square is even.

Finding a proof

The proof should look like this:

Let x be even.

Therefore, x^2 is even.

Let y be a number such that y^2 is even.

Therefore, y is even.

So a number is even if and only if its square is even.
The first half of this proof was an exercise in the last chapter. In the second half of the proof, we begin with,

Let y^2 be even,

and then write this in symbols,

$$y^2 = 2K$$

for some whole number K.

We then look for a reason why y should be even. As 2 divides the right-hand side, it also divides the left-hand side. So 2 divides one (or both) of the factors on the left. These are both y, so 2 divides y.

Proof

The final proof (including both parts) is:

Let x be even,

so that,

$$x = 2K$$

for some whole number K.

Then,

$$\begin{aligned} x^2 &= (2K)^2 \\ &= 4K^2 \\ &= 2(2K^2) \end{aligned}$$

which is even.

Therefore, if x is even then x^2 is even.

Conversely, let y be a whole number such that y^2 is even.

So,

$$y^2 = 2K$$

for some whole number K.

As 2 divides 2K, 2 divides y^2.

So 2 divides either y or y, that is, 2 divides y.

Therefore, if y^2 is even then y is even.

Therefore, a whole number is even if and only if its square is even.

Example 2

Prove that a whole number is divisible by 9 if and only if the sum of its digits is divisible by 9.

Finding a proof

The main problem is to find a way to symbolise the number so that the sum of its digits can also be written in symbols. It will be necessary to have a symbol for each digit. If the number written in the usual decimal form is,

$$a_n a_{n-1} \ldots a_2 a_1 a_0$$

(i.e. a_0 in the units column, a_1 in the tens column, and so on),
then the number itself is,

$$10^n a_n + \ldots + 100a_2 + 10a_1 + a_0$$

and the sum of the digits is,

$$a_n + \ldots + a_2 + a_1 + a_0$$

The reason why one of these is divisible by 9 if and only if the other one is, is that the difference between the two numbers is,

$$\underbrace{99 \ldots 9}_{\text{n nines}}a_n + \ldots + 99a_2 + 9a_1$$

which is obviously divisible by 9. The proof could be set out as below.

Proof

Let x be a number with digits $a_n, a_{n-1}, \ldots, a_2, a_1, a_0$
if x is divisible by 9 then,

$$10^n a_n + \ldots + 100a_2 + 10a_1 + a_0$$

is divisible by 9.

So,

$$(10^n a_n + \ldots + 100a_2 + 10a_1 + a_0) - (\underbrace{99 \ldots 9}_{\text{n nines}}a_n + \ldots + 99a_2 + 9a_1)$$

is divisible by 9,

that is,

$$a_n + \ldots + a_2 + a_1 + a_0$$

is divisible by 9,

that is, the sum of the digits is divisible by 9.

If the sum of the digits is divisible by 9,

$$a_n + \ldots + a_2 + a_1 + a_0$$

is divisible by 9,

so,

$$(a_n + \ldots + a_2 + a_1 + a_0) + (\underbrace{99 \ldots 9}_{\text{n nines}}a_n + \ldots + 99a_2 + 9a_1)$$

is divisible by 9,

that is, the number itself,

$$10^n a_n + \ldots + 100a_2 + 10a_1 + a_0$$

is divisible by 9.

Therefore, a number is divisible by 9 if and only if the sum of its digits is divisible by 9.

Note

In this case, where the steps in the second half are just the reverse of those in the first half, it is possible to combine the steps in a string of, "if and only ifs":
A number with digits $a_n \ldots a_2a_1a_0$ is divisible by 9,

iff $10^n a_n + \ldots + 100a_2 + 10a_1 + a_0$ is divisible by 9,

iff $(10^n a_n + \ldots + 100a_2 + 10a_1 + a_0) - (99 \ldots 9a_n + \ldots + 99a_2 + 9a_1)$ is divisible by 9

iff $a_n + \ldots + a_2 + a_1 + a_0$ is divisible by 9,

iff the sum of the digits is divisible by 9.

However it is rare that this is possible, and it is usually better to keep separate the two parts of an "if and only if" proof.

Exercises

(Gradings of exercises: * easy, ** moderate, *** difficult.)

***1.** Are the following two "iff" statements:

"Something is an A if and only if it is a B"

and,

"Something is a B if and only if it is an A"

equivalent?

***2.** Consider the statement,

"Something is an A if and only if it is a B"

(a) Write down the two assertions made by the above "iff" statement.
(b) How many parts does the proof of the given "iff" statement contain?
(c) Does the proof of the two "all" statements:

"All As are Bs"

and

"All Bs are As"

complete the proof of the given "iff" statement?

***3.** (a) Give one example of a true "iff" statement.
(b) Give one example of a false "iff" statement.

***4.** Are the following "iff" statements true or false?:
(a) "An even number is prime if and only if it is 2".
(b) "An odd number is prime if and only if it is 3".

***5.** One of the following three statements is true. State which one it is.
(a) A figure is a polygon if and only if it is a triangle.
(b) A figure is a circle if and only if it is a polygon.
(c) A figure is a polygon if it is a triangle.

***6.** Prove that a whole number is odd if and only if its square is odd.

***7.** Write the "iff" statement:

x is a non-zero real number if and only if x^2 is positive in:

(a) "necessary and sufficient form.
(b) "if . . . then . . . and conversely" form.

***8.**. Rewrite the "iff" statement,

"Something is an A if and only if it is a B"

in five different forms.

****9.** Prove that a number is divisible by 4 if and only if its last two digits form a number divisible by 4.

****10.** Prove that a triangle is isosceles if and only if two of its angles are equal. (An isosceles triangle is by definition a triangle with two equal sides.)

****11.** Let m and n be two integers. Show that $m^3 - n^3$ is even if and only if $m - n$ is even.

****12.** Show that a triangle has 3 equal sides if and only if it has 3 equal angles.

****13.** Show that a number is divisible by 3 if and only if the sum of its digits is divisible by 3.

****14.** Prove that the lines,

$$Ax + By + E = 0$$

and,

$$Cx + Dy + F = 0$$

are parallel if and only if $AD - BC = 0$

****15.** Prove that the simultaneous equations,

$$Ax + By + E = 0$$

and,

$$Cx + Dy + F = 0$$

have exactly one solution if and only if $AD - BC \neq 0$.

***16. Prove that the lines,

$$Ax + By + E = 0$$

and,

$$Cx + Dy + F = 0$$

are perpendicular if and only if $AC + BD = 0$
(Do not assume without proof that the product of the gradients of two perpendicular lines is -1.)

***17. Prove that three distinct points,

$$(x_1, y_1), (x_2, y_2) \text{ and } (x_3, y_3)$$

are collinear if and only if,

$$(x_2y_3 - x_3y_2) - (x_1y_3 - x_3y_1) + (x_1y_2 - x_2y_1) = 0$$

***18. Prove that a real number is rational if and only if its decimal expansion is terminating or (eventually) repeating.

Linear algebra

**19. Show that a vector in $\mathbf{R}^3$ is a linear combination of (1, 1, 2) and (2, 2, 3) if and only if it lies on the plane, $x = y$.

Calculus

**20. Show that y_1 and y_2 are solutions of,

$$\frac{d^2y}{dx^2} + A\frac{dy}{dx} + By = 0$$

if and only if $y_1 + y_2$ and $y_1 - y_2$ are solutions.

4

"Some" statements

An existential generalisation, or "some" statement, states that there exists something which satisfies a certain condition. For example, "245 is a multiple of 7" says,

"There exists a whole number x such that $245 = 7 \times x$"

or,

"$245 = 7 \times x$ for some whole number x"

The expressions, "there is . . .", and "there are . . .", "there exist . . .", all mean "there is at least one . . .". For example, the statement:

"There is a real number x such that $x^2 = 2$"

is true; the fact that there are two real numbers such that $x^2 = 2$ is not relevant to the truth of the statement.

To prove a "some" statement, the usual method is to exhibit or "construct" a thing satisfying the given condition.

Example 1

Prove that 245 is a multiple of 7.

Proof

We have to prove that,

$$245 = 7 \times x$$

for some x.

But,

$$245 = 7 \times 35$$

so,

$$x = 35$$

is a solution.

Therefore, 245 is a multiple of 7.

The general form of a "some" statement is,

"Some As are Bs"

or

"Some A is a B"

(For example, "Some whole number x is such that 245 $= 7 \times x$".)

As with "all" statements, there are alternative forms, for example,

- Something is both an A and a B.
- There exists something that is both an A and a B.
- There is an A that is also a B.
- There is at least one A that is a B.

There is at least one "some" statement that cannot be put in the form, "Some A are B", namely, "something exists" or, "there is something". It is difficult to see how this statement can be proved; if there is a way, then it is the business of philosophy rather than mathematics. The mathematical attitude to existence is illustrated by the following old joke: A philosopher, a physicist and a mathematician were travelling by train to Victoria. Just after crossing the border they saw a paddock containing one black sheep. "Look at that", said the physicist, "Victorian sheep are black". "But", said the mathematician, "all we've established is that there exists at least one Victorian sheep that is black". The philosopher said, "What we actually *know* is that at least one Victorian sheep is black, on at least one side".

Non-constructive existence proofs, that is, proofs of the existence of things without exhibiting them, are occasionally possible.

Example 2

Show that there is a solution of $f(x) = 0$ between $x = 0$ and $x = 1$, where,

$$f(x) = x^5 - 2x^3 + 5x^2 + x - 2$$

(There is no formula for the roots of a fifth-power equation, so one cannot straightforwardly exhibit or construct the solutions, as one could with a quadratic.)

Proof

$$f(0) = -2$$

which is negative,

while,

$$f(1) = 3$$

which is positive.

So the graph of f(x) must cross the x-axis somewhere between 0 and 1; that is, there is a solution of $f(x) = 0$ somewhere between 0 and 1.

Note

This proof depends on the fact that a continuous function which is negative somewhere and positive somewhere else must be zero at some point in between. Is this obvious enough to use as the basis of a proof? For ordinary purposes, it is generally considered adequate, but work on the foundations of calculus has shown that it is possible to prove such results from ones that are perhaps a little more obvious. (See M. Spivak, *Calculus* (1980), (2nd edn), Publish or Perish, Berkeley, p. 108 or S. R. Lay, *Analysis: An Introduction to Proof* (1986), Prentice Hall, Englewood Cliffs, NJ, p. 159.)

Non-constructive existence proofs such as the above are useful in methods of approximation in computing. Once it has been established that there exists a solution of $x^5 - 2x^3 + 5x^2 + x - 2 = 0$, there is some point in applying a method, such as Newton's method, to approximate the solution. Applying such computational techniques when the solution being sought does not in fact exist can lead to strange outputs that are difficult to interpret.

Generally, constructive existence proofs are preferred to non-constructive ones, when they are possible.

Linear algebra

A sum of scalar multiples of some vectors is called a *linear combination* of them. Thus, a vector $\underset{\sim}{v}$ is a linear combination of the vectors $\underset{\sim}{v}_1, \ldots, \underset{\sim}{v}_n$ if *there exist* scalars $\alpha_1, \ldots, \alpha_n$ such that,

$$\underset{\sim}{v} = \alpha_1\underset{\sim}{v}_1 + \ldots + \alpha_n\underset{\sim}{v}_n$$

Example 3

Show that the vector $(1, 0, -1)$ in $\mathbf{R}^3$ is a linear combination of $(1, 1, 1)$ and $(1, 2, 3)$.

Finding the proof

We want to show there exist real numbers α_1 and α_2 such that,

$$(1, 0, -1) = \alpha_1(1, 1, 1) + \alpha_2(1, 2, 3)$$

so we find α_1 and α_2 by solving these equations:

$$\begin{aligned} \alpha_1 + \alpha_2 &= 1 \\ \alpha_1 + 2\alpha_2 &= 0 \\ \alpha_1 + 3\alpha_2 &= -1 \end{aligned}$$

Solving this system we find,

$$\alpha_1 = 2, \ \alpha_2 = -1$$

Proof

For the proof it is sufficient to simply write:

$$(1, 0, -1) = 2(1, 1, 1) + (-1)(1, 2, 3)$$

so $(1, 0, -1)$ is a linear combination of $(1, 1, 1)$ and $(1, 2, 3)$.

Example 4

The vectors $\underset{\sim}{v}_1, \ldots, \underset{\sim}{v}_n$ are said to be *linearly dependent* if there exist scalars $\alpha_1, \ldots, \alpha_n$, not all zero, such that,

$$\alpha_1\underset{\sim}{v}_1 + \ldots + \alpha_n\underset{\sim}{v}_n = \underset{\sim}{0}$$

Show that the vectors (1, 1, 0), (2, 1, 1) and (7, 6, 1) are linearly dependent.

Proof

We solve,

$$a(1, 1, 0) + b(2, 1, 1) + c(7, 6, 1) = (0, 0, 0)$$

so,

$$\begin{aligned} a + 2b + 7c &= 0 \\ a + b + 6c &= 0 \\ b + c &= 0 \end{aligned}$$

Solving, we find,

$$\begin{aligned} a &= -5c \\ b &= -c \end{aligned}$$

so there are infinitely many solutions.

So there exist a, b, c, not all zero, such that,

$$a(1, 1, 0) + b(2, 1, 1) + c(7, 6, 1) = (0, 0, 0)$$

for example, $a = -5, b = -1, c = 1$.

Therefore the vectors are linearly dependent.

Calculus

Example 5

Show that there exists a (non-zero) function, y of x such that,

$$\frac{d^2y}{dx^2} = -y \quad \text{and} \quad y(0) = y(\pi) = 0$$

Proof

As is usual with differential equations, one finds the general solution of the equation, and then asks how the initial conditions restrict the general solution. The general solution of,

$$\frac{d^2y}{dx^2} = -y$$

is,

$$y = A \sin x + B \cos x$$

Substituting $x = 0$, $y = 0$ gives,

$$0 = A.0 + B.1$$

so,

$$B = 0$$

thus,

$$y = A \sin x$$

This already satisfies the second condition $y(\pi) = 0$, so there is no further restriction. There are, therefore, many functions y satisfying the conditions of the problem, and we need only exhibit one, say $y = \sin x$.

Exercises

(Gradings of exercises: * easy, ** moderate, *** difficult.)

***1.** Consider the "some" statement

"7073 is a multiple of 643"

(a) How would you attempt to prove it?
(b) Prove it.

***2.** Do there exist two integers whose sum is 73 and whose difference is 11? Prove your answer.

***3.** Consider the "some" statement:

"There exists a number x such that, $x^3 + 7x^2 - x - 7 = 0$"

To prove the above statement we can exhibit one x satisfying the condition,

$$x^3 + 7x^2 - x - 7 = 0$$

(a) Verify that $x = -7$ is such a number.
(b) Does the existence of one x satisfying the given condition exclude the possibility of other such xs?
(c) Is the existence of more than one x satisfying the given condition relevant to the truth of the "some" statement?

***4.** A *perfect* number is one which equals the sum of its factors (counting 1 as a factor, but not the number itself). Show that there exists a perfect number.

***5.** Rewrite the "some" statement,

"63 is a multiple of 9"

in four different forms.

***6.** Determine the truth or falsity of the two statements:
(a) For all $x \in \mathbf{R}$, $x^2 \geqslant 0$

(b) For some $x \in \mathbf{R}$, $x^2 > 0$
and then prove your answers.

***7.** Determine the truth or falsity of the following "some" statements:
(a) For some $x \in \mathbf{Z}$, $49 = 7x$
(b) For some $x \in \mathbf{Z}$, $50 = 7x$
(c) For some $x \in \mathbf{R}$, $x^2 = 16$
(d) For some $x \in \mathbf{R}$, $x^2 \geqslant 1000$
(e) For some $x \in \mathbf{R}$, $x^2 = -8$
(f) For some $x \in \mathbf{R}$, $x^3 = -8$
(g) For some $x \in \mathbf{Z}$, $x^3 = -8$
(h) For some $x \in \mathbf{R}$, $x - 7 = x$

***8.** Show that there exists a real number x such that x^2 is irrational but x^4 is rational.

***9.** For each of the following statements, state whether it is true or false and then prove your answer:
(a) For some positive x, both x and $x^2 + 10x + 16$ are prime.
(b) For some positive x, both x and $x^2 + 4x + 3$ are prime.
(c) For some positive x, both x and $x^2 + x + 1$ are prime.
(d) For some positive x, both x and $x^5 + 1$ are prime.

***10.** For each of the following statements state whether it is true or false, then prove your answer:
(a) There exists a positive integer x such that $2^x = x^2$
(b) There exists a positive integer x such that $2^x > x^2$
(c) There exists a positive integer x such that $2^x < x^2$

***11.** Show that there is a solution of, $x^{100} + 5x - 2 = 0$ between $x = 0$ and $x = 1$.

***12.** Translate,

"The line, $y = 2x$ intersects the circle, $x^2 + (y - 2)^2 = 4$"

into a "there exists" statement, and prove it.

****13.** Consider the infinite geometric progression,

$$\frac{1}{2} + \left(\frac{1}{2}\right)^2 + \left(\frac{1}{2}\right)^3 + \ldots + \left(\frac{1}{2}\right)^n + \ldots$$

Prove that there exists an integer N such that the sum of the first N terms of the above series differs from 1 by less than 10^{-6}.

****14.** Show that the curve, $x^2 + xy + y^2 = 1$ has an axis of symmetry.

****15.** Show that there exist at least three different solutions of,

$$88x^3 - 126x^2 + 41x - 2 = 0$$

between $x = 0$ and $x = 1$.

****16.** A formula for e^x is,

$$1 + x + \frac{x^2}{2!} + \frac{x^3}{3!} + \ldots + \frac{x^n}{n!} + \ldots$$

Show that $e^3 = 20.1$, correct to 1 decimal place, by showing:

(a) For any $n \geqslant 4$,

$$\frac{3^n}{n!} < \frac{9}{2}\left(\frac{3}{4}\right)^{n-3}$$

(b) Hence, that there exists N such that,

$$\frac{3^{N+1}}{(N+1)!} + \frac{3^{N+2}}{(N+2)!} + \ldots < \frac{1}{20}$$

(c) Hence, that $e^3 = 20.1$ correct to 1 decimal place.

Linear algebra

***17.** (a) Show that (1, 0, 1) is a linear combination of (0, 0, 1), (0, 1, 1) and (1, 1, 1).

(b) Is (1, 2, 1) a linear combination of (1, 3, 0), (2, 1, 0), (−5, 1, 0) and (1, 1, 0)? Prove your answer.

(c) Show that the vectors (3, 4) and (−36, −48) are linearly dependent.

****18.** Is any linear combination of (1, 1, 0) and (2, 1, 0), other than the zero vector, also a combination of (−1, 0, 1) and (2, 1, 1)?

****19.** Show that two vectors are linearly dependent if and only if one is a scalar multiple of the other.

Calculus

****20.** Does there exist $x < 10{,}000$ such that,

$$\left|\frac{x^2}{1 + x^2}\right| < 0.001?$$

***21.** Are there any differentiable functions which are equal to their second derivative but not their first?

5

Multiple quantifiers

The last three chapters studied "all" statements and "some" statements. "All" and "some" are called *quantifiers* (the universal and the existential quantifier, respectively), since they say something about how many things satisfy some condition. Next we look at statements that contain more than one quantifier. These include many of the important theorems of mathematics (and a few statements of real life as well). They require considerable logical skill, although they do not need any techniques other than those already described for "all" and "some". A number of examples are:

- Every body attracts every other body (two universals).
- Between any two real numbers there is a rational number (two universals and one existential).
- The polynomial $x^2 - 3x + 7$ has at least two real roots, that is, there exists a real number x and there exists a real number y such that $x^2 - 3x + 7 = 0$ and $y^2 - 3y + 7 = 0$ and $x \neq y$ (two existentials).
- Everybody loves somebody sometime (one universal and two existentials).
- There are infinitely many whole numbers, that is, for every number there is a bigger number (one universal and one existential).

The techniques used to prove these are the same as for single "all" and "some" statements.

Example 1

Prove that the square of every even number is a multiple of 4.

(That is, prove that if x is even then $x^2 = 4k$ for some number k.)

This contains one "all" and one "some". The "all" comes first. Something is to be proved about *all* even numbers. So the proof must begin with,

"Let x be an even number."

For each x, we have to show that $x^2 = 4k$ for *some* k, so we must find a k such that $x^2 = 4k$ (of course, we would expect it to be a different k for each x).

Proof

Let x be an even number.

Then, $x = 2y$ for some whole number y.

So,

$$\begin{aligned} x^2 &= (2y)^2 \\ &= 4y^2 \end{aligned}$$

So x^2 is a multiple of 4.

Example 2

One of the most famous proofs in mathematics is Euclid's proof that there are infinitely many prime numbers. The first step of this proof is to express the statement in logical language: For every prime number, there is a larger prime number (one universal and one existential, as in the last example). Thus, the proof begins with,

"Let p be a prime number."

The aim is to find a prime number bigger than p.

Consider one plus the product of all the prime numbers up to p, that is, in symbols, let p_i stand for the ith prime (so $p_1 = 2$, $p_2 = 3$, and so on).

Suppose p is the nth prime, p_n.

Then let,

$$X = p_1 p_2 \ldots p_n + 1$$

X has at least one prime factor q, and q is not any of $p_1, p_2, \ldots, p_n$, since none of these divide X.

So q is a prime different from $p_1, \ldots, p_n$; and since these are the first n primes, q is greater than p_n

This completes the proof.

Notes

1. The explanatory comment, "So $p_1 = 2$, $p_2 = 3$, and so on" is not necessary, but it is usually better to err on the side of more, rather than less, explanation. Reading proofs is not easy.
2. Much the hardest step in the proof is the one introducing the number,

$$X = p_1 p_2 \ldots p_n + 1$$

The point is to produce a number which is not divisible by any of the prime numbers up to p. The genius of the ancient Greek mathematicians in thinking of such steps is the reason why Euclid is still remembered, while his parents are not.
3. The statement "X has a prime factor" is perhaps not completely "obvious", but it is at least "well-known" that every number can be factored into primes. Ideally, this should be proved somewhere earlier.
4. The number X itself is often prime, but not always. For example,

$$2.3.5 + 1 = 31$$

which is prime, but,

$$2.3.5.7.11.13 + 1 = 30031$$

which is 59×509.

5. It is usual to add some statement like "this completes the proof", even in advanced mathematics, as a kind of punctuation to help people read the text quickly. In geometry, the letters Q.E.D. (abbreviation for "quod erat demonstrandum" meaning "which was to be proved") are often added at the end of a proof of a theorem. Some books write the sign ▌ at the end of proofs. Mathematics should always be read through quickly first, to "get the idea", and then read in detail later. It is often impossible to understand the details of a proof until the overall idea or strategy of the proof is understood. Regrettably, not all authors regard it as their job to help their readers by making their strategy clear.
6. What about proving, "There are infinitely many whole numbers"? Notice that this was implicitly assumed in the above proof, since we assumed the number $p_1p_2 \ldots p_n + 1$ could always be formed and that there was no danger of "sailing over the edge" of the whole number system. It has not, in fact, been found possible to prove the statement, "There are infinitely many whole numbers" (or, "For every whole number there is a larger whole number") from anything simpler. There is a problem here, if we wish to continue the practice of trying to prove everything from what is obviously true, since it seems dangerous to regard anything about the infinite as completely obvious. The situation has to be put up with, and there seems little future in doubting that there are infinitely many numbers.

Non-constructive existence proofs again

(The rest of this chapter is a little more advanced and could be left out at a first reading.)

Non-constructive existence proofs are especially common in cases where there is an "all" as well as a "some" in the statement to be proved.

Example 3

Any region in the plane can be divided in half by some vertical line. (See Figure 5.1.)

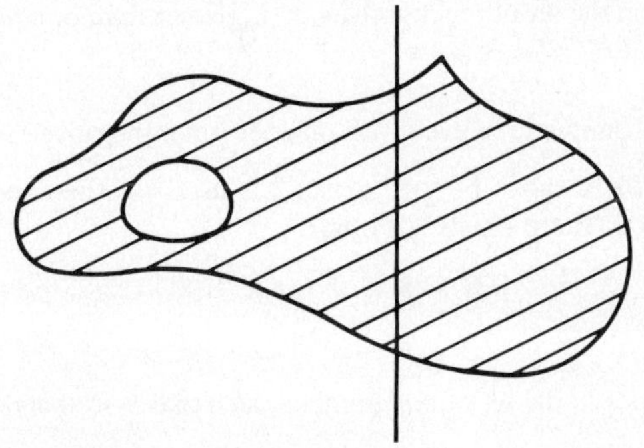

Figure 5.1

Proof

Think of a vertical line moving slowly from the left of the page to the right. As it passes over the region, the amount of shaded area to the left of the line gradually increases from zero to the whole area of the region. So somewhere in between there must be a time when the area to the left is half the whole area. ▌

Notes

1. For simple regions, such as a square or a circle, it is obvious how to construct a vertical line that divides the region in half. For regions of arbitrary shape, however, this proof gives us no method of actually drawing the line (by ruler and compass, for example).
2. The reasoning used in this proof is the same as the reasoning employed in: "The function $f(x) = x^5 + x^3 - 1$ has a zero between 0 and 1, because f(0) is negative and f(1) is positive, which means that the graph of f(x) must cross the x-axis somewhere between 0 and 1." In abstract terms, if a continuous function takes two values, it must take all the values in between as well.

One of the most famous theorems of mathematics, the *Fundamental Theorem of Algebra*, is an existence theorem with a non-constructive proof. It states that any polynomial has a root in the complex numbers. That is, for any polynomial,

$$a_n x^n + a_{n-1} x^{n-1} + \ldots + a_1 x + a_0$$

there exists a complex number y such that,

$$a_n y^n + a_{n-1} y^{n-1} + \ldots + a_1 y + a_0 = 0$$

The proof is difficult. The first convincing proof was given by Gauss in 1799; there has been some debate as to whether his proof was sufficiently rigorous.

Symbols for quantifiers

The quantifiers are sometimes expressed in symbols: The symbol ∀ (an upside-down A, for "all") is used for the universal quantifier. It is read as, "for all", "for every" or "for each". Thus:

$$\forall x \in \mathbf{R} \quad x^2 \geqslant 0$$

is read as,

"For all x in the set of real numbers, x^2 is greater than or equal to 0"

that is, in words,

"The square of every real number is non-negative".

The symbol ∃ (a backwards E, for "exists") is used for the existential quantifier. It is read as, "for some" or "there exists". Thus,

$$\exists x \in \mathbf{R} \quad x < 0 \quad \text{and} \quad x^2 = 2$$

is read as,

"There exists x in the set of real numbers such that $x < 0$ and $x^2 = 2$"

that is,

"The square of some negative real number is 2"

One advantage of using symbols is that it makes the treatment of complicated cases clearer. For example, in statements which have more than one quantifier, it is natural to ask whether changing the order of the quantifiers makes any difference. In fact two universal quantifiers can be interchanged. Thus,

$$\forall x \in \mathbf{R} \ \ \forall y \in \mathbf{R} \quad x^2 - y^2 = (x - y)(x + y)$$

is logically equivalent to,

$$\forall y \in \mathbf{R} \ \ \forall x \in \mathbf{R} \quad x^2 - y^2 = (x - y)(x + y)$$

Two existential quantifiers can also be interchanged,

$$\exists n \in \mathbf{Z} \ \ \exists m \in \mathbf{Z} \quad 7n + 4m = 1$$

and,

$$\exists m \in \mathbf{Z} \ \ \exists n \in \mathbf{Z} \quad 7n + 4m = 1$$

both simply say, "There are two integers, m and n, such that $7n + 4m = 1$".

It is not correct, however, to interchange a universal and an existential. For example,

$$\forall n \in \mathbf{Z} \ \ \exists m \in \mathbf{Z} \quad m > n$$

says that for each integer there is a bigger integer. This is true, but,

$$\exists m \in \mathbf{Z} \ \ \forall n \in \mathbf{Z} \quad m > n$$

says that there is some integer that is bigger than every integer, which is false.

Once this distinction was made, one of the most famous errors in mathematics was corrected. There have been very few cases where one of the great mathematicians claimed to have proved something that later turned out to be false. One of these rare cases occurred in the early 1800s when Cauchy claimed to have proved a statement about the convergence of functions, which was soon found to be false. Cauchy's argument had used some geometrical intuitions about continuity. When these were replaced with more rigorous logical arguments (using the ideas to be described in Chapter 10) it was realised that Cauchy's argument had involved an illegitimate change in the order of quantifiers.

Linear algebra

Example 4

Show that every vector in the XY-plane in $\mathbf{R}^3$ is a linear combination of (1, 1, 0) and (2, 3, 0).

Method

We are asked to show that for *all* vectors $\underset{\sim}{v}$ in the XY-plane, *there exist* scalars a and b such that $\underset{\sim}{v} = a(1, 1, 0) + b(2, 3, 0)$.

Any vector in the XY-plane is of the form (A, B, 0) for some real numbers A and B.

We need to show how to write (A, B, 0) as $a(1, 1, 0) + b(2, 3, 0)$, so we solve,

$$(A, B, 0) = a(1, 1, 0) + b(2, 3, 0)$$

that is, we find a and b in terms of A and B,

$$a + 2b = A$$
$$a + 3b = B$$

Solving, we find $a = 3A - 2B$, $b = B - A$.

So the final proof can be written as follows.

Proof

Any vector in the XY-plane is of the form (A, B, 0) for some $A, B \in \mathbf{R}$.
But,

$$(A, B, 0) = (3A - 2B)(1, 1, 0) + (B - A)(2, 3, 0)$$

so every vector in the XY-plane is a linear combination of (1, 1, 0) and (2, 3, 0).

Example 5

Show that a set of vectors which includes the zero vector is linearly dependent.

Proof

Let $\{\underset{\sim}{v}_1, \ldots, \underset{\sim}{v}_n\}$ be a set of vectors which includes the zero vector. That is, some $\underset{\sim}{v}_i = \underset{\sim}{0}$.
We now need to show a "there exists" statement:

"There exist scalars $a_1, \ldots, a_n$, not all zero, such that $a_1\underset{\sim}{v}_1 + \ldots + a_n\underset{\sim}{v}_n = \underset{\sim}{0}$"

We look for a particular choice of $a_1, \ldots, a_n$ that will do this, remembering that $\underset{\sim}{v}_i = \underset{\sim}{0}$. We can see that one choice is $a_i = 1$, and all other a's zero. All we need to write is:

Then,

$$0\,\underset{\sim}{v}_1 + \ldots + 0\,\underset{\sim}{v}_{i-1} + 1\,\underset{\sim}{v}_i + 0\,\underset{\sim}{v}_{i+1} + \ldots + 0\,\underset{\sim}{v}_n = \underset{\sim}{0}$$

so, $\{\underset{\sim}{v}_1, \ldots, \underset{\sim}{v}_n\}$ is linearly dependent.

(The converse of this theorem is not true—if a set of vectors is dependent, it does not follow that one of them must be zero.)

Calculus

An important use of logic is in making concepts precise through definitions. We have already seen two examples:

- "x is a multiple of y" means, "There exists z such that $x = yz$";
- "There are infinitely many primes" means, "For any prime there is a larger prime".

Such definitions not only improve precision in thinking, but prepare the ground for proofs. Logical definitions using many quantifiers are especially important in calculus. This will be treated fully in Chapter 10, but in the meantime we give a short example.

Example 6

It is well known that, "exponentials grow faster than powers of x". For example, although 2^x is less than x^{100} for many small values of x, 2^x grows faster and eventually overtakes x^{100}. (See Figure 5.2.) How can we make this statement precise?

First, the statement is about *all* exponentials and *all* powers of x. So the statement will begin:

"For all a and for all n, . . . a^x . . . x^n . . ."

The next step is to translate the "eventually" in, "a^x is eventually greater than x^n, and stays greater after that" (the x would be different for different a and n).

The final expression is:

"For all $a > 1$ and for all positive integers n there exists x_0 such that for all $x > x_0$, $a^x > x^n$"

Note

Note that the, "for all $x > x_0$" can be placed at the end of the statement without changing its meaning, but the, "there exists y" must remain where it is, between the second and third "all". "For all $a > 1$" and "For all positive integers n" are interchangeable.

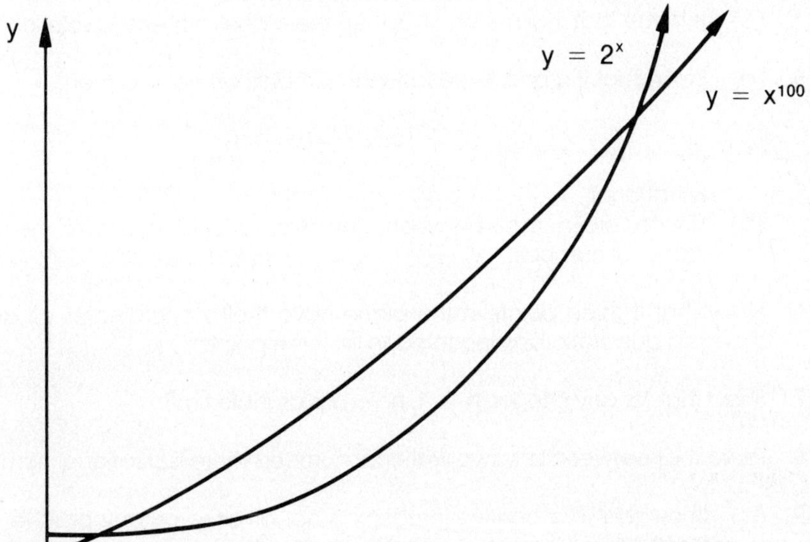

Figure 5.2

Exercises

(Gradings of exercises: * easy, ** moderate, *** difficult.)

*1. (a) How many quantifiers are there in the sentence,

"For some integers m and n, $3m + 2n = 1$"

(b) Decide whether the given statement is true or not and then prove your answer.

**2. Let $f(x) = ax^2 + bx + c$ where $a, b, c \in \mathbf{R}$.
Show that if either a or b is non-zero and if $b^2 - 4ac \geqslant 0$ then there exists $y \in \mathbf{R}$ such that $f(y) = 0$.

**3. Show that for any whole number n there exists a whole number N such that,

$$\frac{1}{N} - \frac{1}{N+1} < \frac{1}{n}$$

**4. (a) Show that for any real numbers a, b, c, d, e, f, if,

$$ad - bc \neq 0$$

then there exists a solution of the simultaneous equations:

$$ax + by = e$$
$$cx + dy = f.$$

(b) Is it true that if there exists a solution then $ad - bc \neq 0$?
(c) Is it true that if $ad - bc = 0$, then there does not exist a solution?

**5. (a) Prove that if a and b are rational numbers with $a \neq b$ then,

$$a + \frac{1}{\sqrt{2}}(b - a)$$

is irrational.
(b) Hence prove that between any two rational numbers there is an irrational number.

**6. Show that if three points in the plane have their x-coordinates all different, there is a quadratic passing through all three points.

**7. Show that for any integer $n \geqslant 1$, $n^5 - n$ is divisible by 5.

***8. Prove that between any two irrational numbers there is a rational number.

***9. (a) Show that if a positive integer x is not divisible by any positive integer less than or equal to $\sqrt{x}$ then it is a prime number.
(b) Is the statement still true if "or equal to" is omitted?

***10. Prove that if a plane figure has an axis of symmetry, and if a point on the axis is a centre of symmetry, then the figure has another axis of symmetry. (The terminology is explained in Chapter 2, Exercise 28.)

***11. Every one of six points is joined to every other one by either a red or a blue line. Show that there exist three of the points joined by lines of the same colour.

Linear algebra

**12. Show that any vector $(x, y, z) \in \mathbf{R}^3$ is a linear combination of $(1, 0, 0)$, $(1, 1, 0)$ and $(1, 1, 1)$.

**13. (a) Show that a set of vectors, one of which is a scalar multiple of another, is linearly dependent.
(b) Show that a set of vectors, one of which is the sum of two of the others, is linearly dependent.

***14. Show that the vectors $\{\underset{\sim}{v}_1, \ldots, \underset{\sim}{v}_n\}$ are linearly dependent if and only if, for some i, $\underset{\sim}{v}_i$ is a linear combination of $\underset{\sim}{v}_1, \ldots, \underset{\sim}{v}_{i-1}$.

Calculus

*15. Express in logical terms: n! grows faster than 2^n.

***16. Does there exist a function, other than the zero function, such that,

$$f(0) = f'(0) = f''(0) = \ldots = f^{(n)}(0) = \ldots = 0?$$

6

"Not", contradiction and counterexample

"Not" and counterexamples

If "not" is put in front of a statement it negates the statement. For any statement p, either "p", or "not-p" is true and the other is false. "Not-p" can also be expressed as "p is false". It is sometimes called the *negation* (or *contradictory*) of p. For example, "$\sqrt{2}$ is irrational" means the same as "not-($\sqrt{2}$ is rational)", that is, in everyday English, "$\sqrt{2}$ is not rational". This statement is the negation of "$\sqrt{2}$ is rational".

We next consider the negations of "all" and "some" statements. The statement "all animals are horses" is false (so its negation, "not all animals are horses", is true). The reason the first statement is false is that there are some animals that are not horses.

In general, the negation of "All As are Bs" is logically equivalent to the statement "Some A is not a B". If "Not all As are Bs" is true, then "Some A is not a B" is true, and vice versa; if one is false, the other is false. Therefore, to show that a "not all" statement is true, or to show that an "all" statement is false, it is necessary to find one *counterexample*.

Example 1

Is "All multiples of 3 are multiples of 6" true or false? Prove your answer.

Answer

False, because 9 is a multiple of 3, but is not a multiple of 6. ▌

(9 is called a "counterexample" to the "all" statement "All multiples of 3 are multiples of 6".)

Example 2

Is "All prime numbers are odd" true or false? Prove your answer.

Answer

False, because 2 is a prime number and it is not odd. ▌

The above examples demonstrate that showing "All As are Bs" is false, requires a quite different strategy to that required to show that the statement is true. To prove, "All As are Bs" is true, we must show something about *every* A, and so the proof begins with "Let x be an A" and must derive "x is a B" from this. But to show "All As are Bs" is *false*, all we need to do is exhibit *one* counterexample, that is, a single A that is not a B.

Usually an "all" statement that is false has many counterexamples, but when proving it is false, it is considered stylish to exhibit just one (as in the first example above.)

"Not all As are Bs" is definitely not equivalent to, "All As are not Bs", that is, "No As are Bs". For example, "Not all animals are horses" is true, (since there are animals that are not horses, such as sloths), while "No animals are horses" is false (since some animals *are* horses). This is illustrated clearly in the Venn diagram in Figure 6.1.

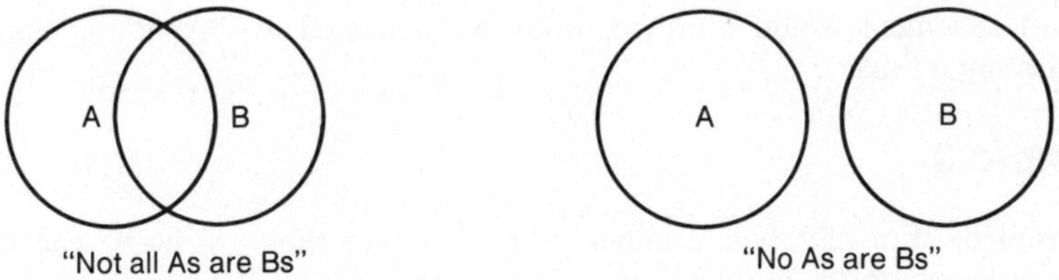

Figure 6.1

(It should be noted that occasionally in everyday English "All As are not Bs" *is* used to mean "Not all As are Bs", as in "All is not well" and "All that glitters isn't gold".)

The contrapositive

It is sometimes natural to introduce "Not" into a proof to make the proof easier. In Chapter 3 we proved the "all" statement, "For all whole numbers y, if y^2 is even then y is even", using a direct proof. We could have, however, argued equally well that, "If y is *not* even, it is odd. As we can show that the square of an odd number is odd, not even, then, if the square of y is even, y itself cannot be odd. So y is even." Such an argument illustrates the following important logical equivalence. The "all" statement,

"All As are Bs"

is logically equivalent to the statement, called its *contrapositive*,

"All not-Bs are not-As"

(or, "Anything that is not a B is not an A")

(Note carefully the opposite order of B and A in the two statements.)

The equivalence of an "all" statement and its contrapositive is clear from a Venn diagram. If "All As are Bs" is true, the set of As is inside the set of Bs. (See Figure 6.2.) Anything that is not a B is outside the set of Bs, and hence is outside the set of As as well (i.e. is not an A).

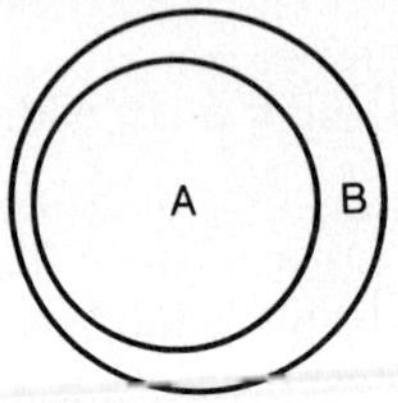

Figure 6.2

(One could also prove the "if and only if" statement, "All As are Bs if and only if all not-Bs are not-As", using the techniques of Chapter 3.) For example, all horses are animals, so the set of horses lies inside the set of animals. Therefore, whatever is not an animal is not a horse. So,

"All horses are animals"

is logically equivalent to,

"All not-animals are not-horses"

This last statement would more naturally be expressed as, "Anything that is not an animal is not a horse".

Example 3

The proof of "For all whole numbers y, if y^2 is even then y is even" can therefore be written as shown in the proof below.

Proof

Let x be an odd whole number.

Then,

$$x = 2k + 1$$

for some whole number k.

So,

$$\begin{aligned} x^2 &= (2k + 1)^2 \\ &= 4k^2 + 4k + 1 \\ &= 2(2k^2 + 2k) + 1 \end{aligned}$$

So x^2 is odd.

We have proved that if x is odd then x^2 is odd. Therefore, if x^2 is not odd then x is not odd, since this statement is the contrapositive of the last one. Therefore, if x^2 is even then x is even.

Proof by contradiction

Proof by contradiction, or *reductio ad absurdum* proof, works by assuming the negation of the proposition to be proved and deducing a contradiction. This shows the negation is false, and hence that the original proposition is true. Example 4 is the classic proof of this kind.

Example 4

Prove that $\sqrt{2}$ is irrational.

Finding the proof

A number is rational when it is of the form m/n for some integers m and n. So we are asked to show that for all integers m and n, m/n is not $\sqrt{2}$. The most natural thing to do is to assume that m/n *is* $\sqrt{2}$, for some integers m and n, and deduce a contradiction.

The first line should be:

Suppose,

$$\frac{m}{n} = \sqrt{2}$$

for some integers m and n.

By cancelling out any common factors, we can suppose m and n have no common factor. (This is often done with fractions, but the reason for doing it here is only clear later in the proof.)

The only possible way to proceed is to apply the definition of $\sqrt{2}$: It is the number whose square is 2. So we square both sides:

$$\frac{m^2}{n^2} = 2$$

The equation becomes simpler if we multiply by n^2:

$$m^2 = 2n^2$$

At this stage some new idea is needed. Since we are dealing with integers, it may be useful to look at divisibility: The right-hand side is even (divisible by 2) so the left-hand side is too. So,

$$m^2 \text{ is even}$$

Therefore,

$$m \text{ is even}$$

So, $m = 2k$ for some integer k.

Putting this into the equation,

$$m^2 = 2n^2$$

gives,

$$(2k)^2 = 2n^2$$
$$4k^2 = 2n^2$$
$$2k^2 = n^2$$

Therefore, n^2 is even.

So n is even.

We have shown that m and n are both even, that is, they have 2 as a common factor. This contradicts the original assumption that m and n have no common factor.

Therefore $\sqrt{2}$ is not rational. The final version of this proof would normally be written as shown below:

Proof

Suppose,

$$\frac{m}{n} = \sqrt{2}$$

for some integers m, n.

By cancelling any common factors, we may suppose m and n have no common factor.

Then,

$$\frac{m^2}{n^2} = 2$$
$$m^2 = 2n^2$$

So m^2 is even and thus m is even.

So,

$$m = 2k$$

for some integer k.

So,

$$(2k)^2 = 2n^2$$
$$4k^2 = 2n^2$$
$$2k^2 = n^2$$

So n^2 is even and thus n is even.

So m and n are even, contradicting the assumption that m and n have no common factor.

So $\sqrt{2}$ is not of the form m/n, so $\sqrt{2}$ is irrational. ▌

This is a very neat proof, but it is important to understand that a large part of the thinking happens before the first line is written. This fact is covered up in the final version.

In principle, any proof that a number is irrational can be expected to be a *reductio ad absurdum*. The irrationality of e is not difficult to prove; a proof can be found in M. Spivak, *Calculus* (1980), (2nd edn) p. 353 or in R. Courant and H. Robbins, *What is Mathematics?* (1941), Oxford University Press, London, pp. 298–9. Proving the irrationality of π is however more difficult. Although π has been well-known since ancient times, its irrationality was only proved (by Lambert) around 1761. A proof can be found in M. Spivak, *Calculus* (1980), (2nd edn), Publish or Perish, Berkeley, Chap. 16.

Sometimes there is a choice about whether a proof is to be treated as a *reductio ad absurdum* or not. For example, the proof that there are infinitely many prime numbers could begin: "Suppose there is a largest prime number", and deduce a contradiction, by showing there is a larger one. Usually *reductio ad absurdum* arguments are avoided where there is a choice, since a string of statements that are all false is harder for the mind to deal with than a string of true statements, which can be understood.

"Not" and multiple quantifiers (more advanced)

At the beginning of this chapter we saw that the contradictory of an "all" statement is a "some . . . not" statement. This can be stated more formally as:

If Px is some statement involving x,

$$\text{Not } \forall x \text{ Px}$$

is logically equivalent to,

$$\exists x \text{ Not Px}$$

That is, to say that "Not all x are P" is the same as to say that "Some x is not P". Thus, "Not all animals are horses" is equivalent to "Some animal is not a horse".

Similarly,

$$\text{Not } \exists x \text{ Px}$$

is logically equivalent to,

$$\forall x \text{ Not Px}$$

These patterns can in fact be extended to negate (i.e. find the contradictory of) statements with multiple quantifiers. Consider,

$$\text{Not } \forall x \ \exists y \text{ Qxy}$$

where Qxy is some statement involving x and y. First, "not all" is "some . . . not", so this is equivalent to,

$$\exists x \text{ Not } \exists y \text{ Qxy}$$

Then, "not . . . some" is "all . . . not", so this is equivalent to,

$$\exists x \ \forall y \text{ Not Qxy}$$

Example 5

Is it true that,

$$\forall x \in \mathbf{R} \quad \exists y \in \mathbf{R} \quad y^2 = x$$

(i.e. does every real number have a real square root?)

Prove your answer.

Proof

The statement is not true.
A counterexample is,

$$x = -3$$

Because -3 is not the square of any real number, as the square of any real number is positive, this proves,

$$\forall y \in \mathbf{R} \quad y^2 \neq -3$$

So,

$$\exists x \in \mathbf{R} \quad \forall y \in \mathbf{R} \quad y^2 \neq x$$

This is equivalent to,

$$\text{Not} \quad \forall x \in \mathbf{R} \;\; \exists y \in \mathbf{R} \;\; y^2 = x$$

which we had to show.

Note

It is intuitively obvious that to show that "Every real number has a real square root" is false, we must find a number that does not have a real square root. The use of symbols to make this intuition explicit helps us to understand it better and allows us to generalise to more complex cases. A general rule to find the contradictory of a statement with any number of quantifiers is: Change each $\forall$ to $\exists$, and vice versa, and put the "not" after the quantifiers. For example,

$$\text{Not} \quad \forall x \quad \exists y \quad \forall z \quad Rxyz$$

is equivalent to,

$$\exists x \quad \forall y \quad \exists z \quad \text{Not} \quad Rxyz$$

Linear algebra

Example 6

Show that (1, 0, 1) is not a linear combination of (2, 3, 0) and (1, 1, 0).

Proof 1

If (1, 0, 1) were a linear combination of (2, 3, 0) and (1, 1, 0), there would exist scalars a and b such that,

$$(1, 0, 1) = a(2, 3, 0) + b(1, 1, 0)$$

So,

$$\begin{aligned} 2a + b &= 1 \\ 3a + b &= 0 \\ 0a + 0b &= 1 \end{aligned}$$

The last equation is $0 = 1$, which is a contradiction. So we have shown (using proof by contradiction) that (1, 0, 1) is not a linear combination of (2, 3, 0) and (1, 1, 0). ▮

Proof 2

All linear combinations of (2, 3, 0) and (1, 1, 0) have last co-ordinate zero; (1, 0, 1) does not, so it cannot be a linear combination of (2, 3, 0) and (1, 1, 0). ▮

Recall that vectors $\underset{\sim}{v}_1, \ldots, \underset{\sim}{v}_n$ are *linearly dependent* if there exist scalars $a_1, \ldots, a_n$, not all zero, such that,

$$a_1\underset{\sim}{v}_1 + \ldots + a_n\underset{\sim}{v}_n = \underset{\sim}{0}$$

"Linearly independent" means, "not linearly dependent". Thus vectors are linearly independent when the *only* solution of the equation,

$$a_1\underset{\sim}{v}_1 + \ldots + a_n\underset{\sim}{v}_n = \underset{\sim}{0}$$

is,

$$a_1 = \ldots = a_n = 0$$

Example 7

Show that if $\underset{\sim}{u}$ and $\underset{\sim}{v}$ are linearly independent vectors, then $\underset{\sim}{u} + \underset{\sim}{v}$ and $\underset{\sim}{u} - \underset{\sim}{v}$ are also independent.

Proof

Suppose $\underset{\sim}{u}$ and $\underset{\sim}{v}$ are independent.

We wish to show $\underset{\sim}{u} + \underset{\sim}{v}$ and $\underset{\sim}{u} - \underset{\sim}{v}$ are independent.

Let,

$$a(\underset{\sim}{u} + \underset{\sim}{v}) + b(\underset{\sim}{u} - \underset{\sim}{v}) = \underset{\sim}{0}$$

So,

$$(a + b)\underset{\sim}{u} + (a - b)\underset{\sim}{v} = \underset{\sim}{0}$$

But, since $\underset{\sim}{u}, \underset{\sim}{v}$ are independent, the only solution of,

$$A\underset{\sim}{u} + B\underset{\sim}{v} = \underset{\sim}{0}$$

is,

$$A = B = 0$$

Thus,

$$a + b = 0$$
$$a - b = 0$$

Solving this system, we find $a = 0$ and $b = 0$.

Therefore, we have shown that the only solution of,

$$a(\underset{\sim}{u} + \underset{\sim}{v}) + b(\underset{\sim}{u} - \underset{\sim}{v}) = \underset{\sim}{0}$$

is,

$$a = b = 0$$

So $\underset{\sim}{u} + \underset{\sim}{v}$ and $\underset{\sim}{u} - \underset{\sim}{v}$ are linearly independent. ▌

Calculus

Example 8

Is it true that if a function f is increasing everywhere (i.e. if $x > y$ then $f(x) > f(y)$), then its derivative is always greater than zero?

Answer

This is not true. A counterexample is,

$$f(x) = x^3$$

It is increasing everywhere, but its derivative at zero is zero. ▌

Exercises

(Gradings of exercises: * easy, ** moderate, *** difficult.)

***1.** What is the common feature in showing:
(i) a "'not all" statement to be true?
(ii) an "all" statement to be false?
Give an example in each case.

***2.** You are told that the statement, "Not all As are Bs", is true.
(a) What can you say about the truth or falsity of the statement, "Some A is not a B"?
(b) Give an example.

***3.** What is the contradictory of each of the following statements?
(a) All cars are red.
(b) All men are humans.

***4.** (a) If a statement p is false what can you say about not-p?
(b) If a statement p is true what can you say about not-p?

***5.** (a) Write down the contradictory of the following statement, "My car is red".
(b) Write down the contradictory of the following "all" statement, "All cars are red", in:
(i) "not-all" form.
(ii) "some" form.
(c) Write down the contradictory of the "some" statement, "For some $x \in \mathbf{R}$, $x^2 = 2$"

***6.** Disprove, by counterexample, each of the following statements:
(a) "For all integers m and n, $m + n$ is positive".
(b) "If p and q are irrational numbers then $p + q$ is irrational".
(c) "For all $x \in \mathbf{R}$,

$$\frac{1}{x(x+1)} = \frac{1}{x} + \frac{1}{x+1}$$

***7.** Give an example to show that, "Not all A are B" and, "Some A are B" can both be true.

***8.** Is the following statement true or false? Prove your answer.

"For all real numbers a, b, c, if $ac = bc$ then $a = b$"

***9.** Comment on the reasoning: "One dog has four legs, and no dog has five legs. Therefore one dog has nine legs".

***10.** Use proof by counterexample to prove the truth or falsity of the following statements:
(a) "For all $x \in \mathbf{R}$, $x^2 > 0$"
(b) "Not all multiples of 6 are multiples of 9".

***11.** Are the following statements true or false? Prove your answers.
(a) Let a, b, c be three integers. If a divides c and b divides c, then either a divides b or b divides a.
(b) If a, b, c, d are real numbers with $a < b$ and $c < d$, then $ac < bd$.

****12.** (a) Explain, by listing the steps, how proof by contradiction (*reductio ad absurdum*) works.
(b) What steps would you use to prove by contradiction that $\sqrt{3}$ is irrational?
(c) Carry out the proof that $\sqrt{3}$ is irrational.

****13.** (a) Write down the contradictory of the "all" statement, "All As are not Bs".
(b) What would a direct proof of the statement, "All As are not Bs", look like?
(c) Often it is hard to prove, "All As are not Bs" by proceeding as for any universal. Instead we resort to proof by contradiction. Explain how proof by contradiction works in proving, "All As are not Bs".
(d) Apply this to the proof of "If x^2 is even, then x is not odd".

****14.** Show that no points of the form $(\cos\theta, \sin\theta)$ lie inside the triangle with vertices (0, 0), (0, 0.9) and (0.9, 0).

****15.** Prove that the square of every number is not of the form $4k + 2$ or $4k + 3$. Hence show that no number of the form 100 . . . 003 is a square. (A direct proof is suggested.)

****16.** Prove that each of the following is irrational:
(a) $\sqrt[3]{4}$
(b) $1 + \sqrt{3}$
(c) $\sqrt{3} - \sqrt{2}$

****17.** Show that if a, b, c and d are rational and $\sqrt{b}$ and $\sqrt{d}$ are irrational and,

$$a + \sqrt{b} = c + \sqrt{d}$$

then,

$$a = c \quad \text{and} \quad b = d$$

****18.** Use proof by contradiction to show that the set of primes is infinite.

****19.** Show that if n is a whole number, then $2^n - 1$ and $2^n + 1$ are usually not both prime. What is/are the exceptional case(s)?

****20.** Show that if $2^n - 1$ is prime, then n is prime.
Hint: Consider the contrapositive and use the formula,

$$x^n - y^n = (x - y)(x^{n-1} + x^{n-2}y + \ldots + y^{n-1})$$

****21.** Show that if the product of two whole numbers is even, then at least one of the numbers is even.

****22.** Show that the product of three consecutive positive integers is never a cube.

****23.** Is "All As are not Bs" logically equivalent to "All Bs are not As"? Explain.

****24.** Show that there do not exist three consecutive whole numbers such that the cube of the greatest equals the sum of the cubes of the other two.

****25.** Show that if m and b are real numbers with $m \neq 0$, and the function f is defined by,

$$f(x) = mx + b$$

then if $x \neq y$, $f(x) \neq f(y)$.

****26.** The points A, B, C, D are the midpoints of the sides of a square. Show that not all the angles in the octagon in the middle are equal. See Figure 6.3.

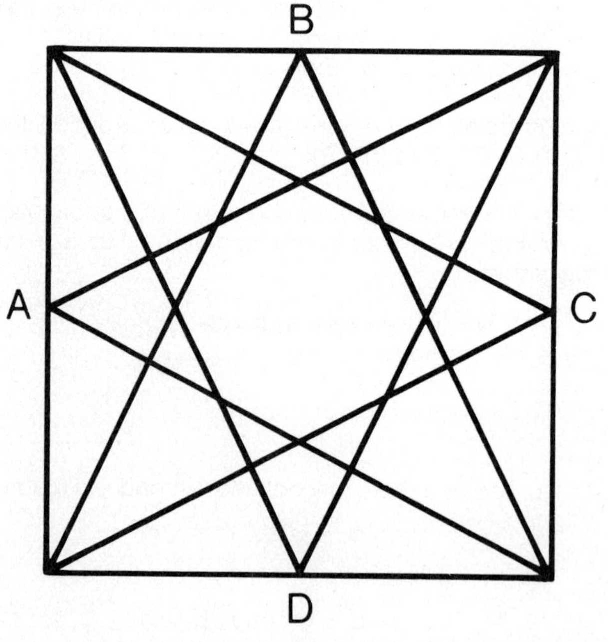

Figure 6.3

***27. In plane geometry, it can be proved that the exterior angle of a triangle is equal to the sum of the two opposite interior angles. That is, in Figure 6.4, $<C = <A + <B$.

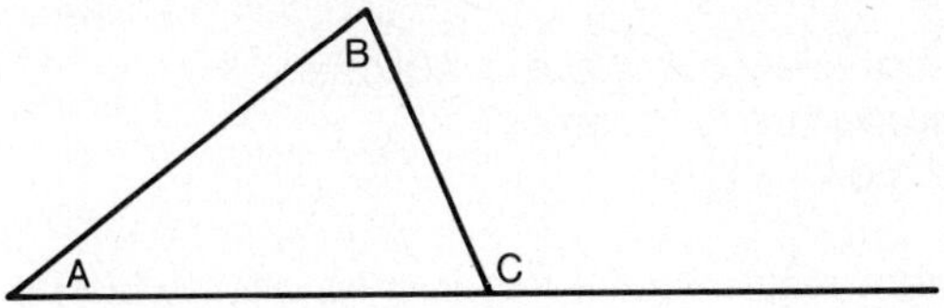

Figure 6.4

Prove by contradiction, using this fact, that if a line cuts two lines such that the alternate interior angles are equal (angles E and F in Figure 6.5) then the two lines are parallel (i.e. they do not meet).

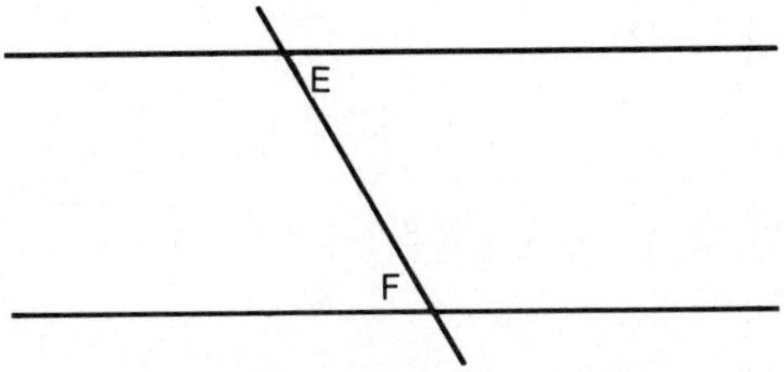

Figure 6.5

***28. Prove that if three points in the plane are not collinear, there exists one, and only one, circle passing through them.

Linear algebra

*29. Show that the vectors (1, 1, 1), (0, 1, 1) and (0, 0, 1) are linearly independent.

**30. Is it true that, if $\underset{\sim}{v}_n$ is not a linear combination of $\underset{\sim}{v}_1, \ldots, \underset{\sim}{v}_{n-1}$, then $\underset{\sim}{v}_1, \ldots, \underset{\sim}{v}_n$ are linearly independent?

**31. If $\underset{\sim}{u}$ and $\underset{\sim}{v}$ are independent, do there exist scalars a and b such that $\underset{\sim}{u} + a\underset{\sim}{v}$ and $\underset{\sim}{u} + b\underset{\sim}{v}$ are dependent?

**32. If $\underset{\sim}{u}$ and $\underset{\sim}{v}$ are independent, do there exist scalars a, b, c, d, all non-zero, such that $a\underset{\sim}{u} + b\underset{\sim}{v}$ and $c\underset{\sim}{u} + d\underset{\sim}{v}$ are dependent?

***33. Prove that a set of vectors $\underset{\sim}{v}_1, \ldots, \underset{\sim}{v}_n$ is linearly independent (i.e. the only solution of $a_1\underset{\sim}{v}_1 + \ldots + a_n\underset{\sim}{v}_n = 0$ is $a_1 = \ldots = 0$), if and only if none of the vectors is a linear combination of the others.

Calculus

*34. Is it true that every continuous function is differentiable?

*35. Is it true that if the derivative of a function is always positive, then its graph crosses the x-axis?

****36.** (a) Is it true that if $f(x)$ has a minimum at $x = a$, then $f'(a) = 0$ and $f''(a) > 0$?
(b) Is it true that if $f(x)$ is differentiable and has a minimum at $x = a$, then $f'(a) = 0$ and $f''(a) > 0$?

****37.** (a) Show that every cubic has a real root.
(b) Is the same true for quartics?
(c) Generalise.

****38.** Rolle's theorem states that if $f(x)$ is a function which is continuous on the closed interval $[a, b]$ and differentiable on the open interval (a, b), and if $f(a) = f(b)$, then there exists c between a and b such that $f'(c) = 0$.
(a) Show that the theorem is not true if "$[a, b]$" is replaced by "(a, b)".
(b) Show that the theorem is not true if the condition "differentiable on the open interval (a, b)" is removed.

7

Sets

The language of sets has proved very convenient for expressing a wide variety of mathematical statements. It also aids mathematical thinking by allowing one to think of a collection of related things as a single entity, or set, which can then be given a name or symbol. (The mathematics student will already appreciate the importance of symbols.) The things in a set are called *members*, or *elements*. Sets may be defined by listing the elements in them. For example, $\{1, \pi, 107\}$ is the set consisting of the three numbers: 1, π, and 107. (The order in which the elements are listed is immaterial, so that $\{1, \pi, 107\} = \{107, 1, \pi\}$.) Almost always, however, we collect things together for a reason, that is, because they have some property in common. One speaks of "The set of Fs", meaning the set of all things which have the property F. This set is denoted by,

$$\{x \mid Fx\} \quad \text{or} \quad \{x : Fx\}$$

and read as,

"The set of xs such that Fx"

(There is no significance in choosing x; $\{y \mid Fy\}$ is the same set as $\{x \mid Fx\}$.)

Some sets have special names:

N is the set of natural numbers
Z is the set of integers
Q is the set of rational numbers
R is the set of real numbers
C is the set of complex numbers.

The symbol for "is a member of" is $\in$. Thus,

$$13 \in \{x \mid x \text{ is prime and } 10 < x < 20\}$$

So, "$y \in \{x \mid Fx\}$" is logically equivalent to "Fy".

Two sets are defined as the same, or equal, if they have the same members. In this respect, sets are unlike properties. For example, "having 3 angles equal" and, "having 3 sides equal" are different properties of triangles, but the sets,

$$\{x \mid x \text{ is a triangle with 3 equal angles}\}$$

and,

$$\{x \mid x \text{ is a triangle with 3 equal sides}\}$$

are the same set, since a triangle has 3 equal sides if and only if it has 3 equal angles. Thus for sets A and B, "A = B" means "x ∈ A if and only if x ∈ B". Similarly, "A ⊂ B" (read as, "A is a subset of B") means that everything in A is also in B, that is, "if x ∈ A, then x ∈ B" or "All members of A are members of B" (the case A = B is not excluded).

Often, we define a subset of a given set by choosing those members that have some common property. Thus,

$$\{x \in \mathbf{R} \mid x^2 = 2\},$$

which is read as, "The set of x in **R** such that $x^2 = 2$", is the set of real numbers whose square is 2, that is,

$$\{\sqrt{2}, -\sqrt{2}\}.$$

Relations between sets are sometimes depicted by *Venn diagrams* (called after the nineteenth-century English logician, clergyman and historian John Venn, although actually invented by Leibniz). The elements of a set A are represented by points inside a circle. (If there are many sets, figures other than circles may be needed.) The points outside the circle represent those things not in the set A. (Figure 7.1.) The set of things not in set A is called the *complement* of A, and is denoted by $\bar{A}$ or $\tilde{A}$.

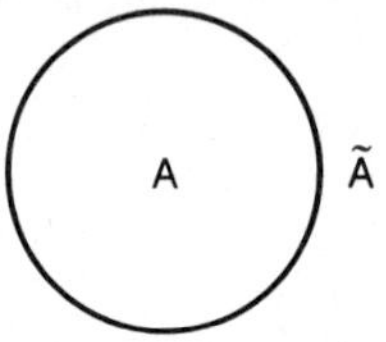

Figure 7.1

For any two sets A and B, the *intersection*, A ∩ B, is the set of things that are in both A and B. (See Figure 7.2.) Thus,

$$A \cap B = \{x : x \in A \text{ and } x \in B\}$$

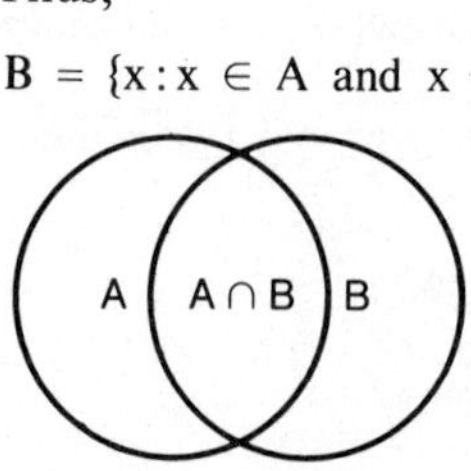

Figure 7.2

The *union* A ∪ B is the set of things which are in either A or B. (See Figure 7.3.) Thus,

$$A \cup B = \{x : x \in A \text{ or } x \in B\}$$

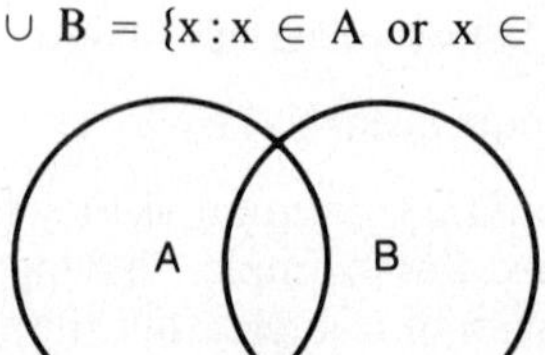

Figure 7.3 A ∪ B is the shaded region

Note that "or" is always used inclusively in mathematics, that is, to mean "and/or". Thus, "x is an F or x is a G" is true when x is an F, or a G, or both.

A ⊂ B is represented in Figure 7.4.

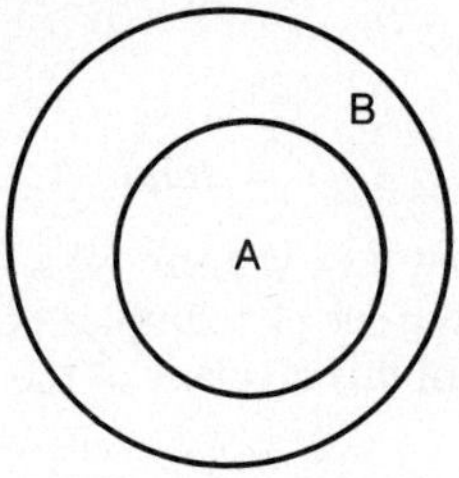

Figure 7.4

This diagram helps to clarify the reason why "All As are Bs" is logically equivalent to its contrapositive, "All not-Bs are not-As". If all points in A are in B, then obviously all points outside B are outside A, and vice versa. Any "all" statement can be represented by this diagram, since "All As are Bs" is logically equivalent to "The set of all As is a subset of the set of all Bs". Venn diagrams are not usually used in representing "some" statements.

Since "A ⊂ B" means, "All members of A are members of B" or, "If x ∈ A then x ∈ B", the proof of a statement of the form "A ⊂ B" must look like this:

Let x ∈ A.

Therefore, x ∈ B.

Similarly, proving "A = B" involves proving an "if and only if" statement, so the proof of "A = B" will usually look like this:

Let x ∈ A.

Therefore x ∈ B.

Then, let x ∈ B.

Therefore, x ∈ A.

Example 1

Prove that the set of multiples of 4 is a subset of the set of even numbers.

Proof

Let x be an element of the set of multiples of 4.

That is, x is a multiple of 4.

So,

$$x = 4k$$

for some integer k,

$$= 2(2k)$$

Therefore, x is a multiple of 2, that is, x is even. So x is in the set of even numbers. This has shown that any member of the set of multiples of 4 is a member of the set of even numbers. So the set of multiples of 4 is a subset of the set of even numbers. ▮

Note

The example is artificially simple, in that the result to be proved does not really say any more than, "All multiples of 4 are even"; the reference to sets adds little in this case. But the example is useful for illustrating something important that happens in general: when proving an "$A \subset B$" statement using,

Let $x \in A$.

Therefore, $x \in B$.

The line after "Let $x \in A$" will usually be an explanation of what "$x \in A$" means. In the above case, "x is in the set of multiples of 4" means just "x is a multiple of 4".

The next example shows a general result about sets, rather than a result about a particular set (such as the set of even numbers). Nevertheless, the techniques of proof are the same as those described above.

Example 2

Prove that $(A \cap B) \cap C = A \cap (B \cap C)$ for any sets A, B, C.

Proof

Let,

$$x \in (A \cap B) \cap C$$

So,

$$x \in A \cap B \quad \text{and} \quad x \in C$$

So,

$$x \in A \quad \text{and} \quad x \in B \quad \text{and} \quad x \in C$$

So,

$$x \in A \quad \text{and} \quad x \in B \cap C$$

So,

$$x \in A \cap (B \cap C)$$

Therefore,

$$(A \cap B) \cap C \subset A \cap (B \cap C)$$

Now let,

$$x \in A \cap (B \cap C)$$

So,

$$x \in A \quad \text{and} \quad x \in B \cap C$$

So,

$$x \in A \quad \text{and} \quad x \in B \quad \text{and} \quad x \in C$$

So,

$$x \in A \cap B \quad \text{and} \quad x \in C$$

So,

$$x \in (A \cap B) \cap C$$

Therefore,

$$A \cap (B \cap C) \subset (A \cap B) \cap C$$

So we have shown that,

$$(A \cap B) \cap C = A \cap (B \cap C) \quad \blacksquare$$

Diagrams are not considered adequate as a proof of such results, though they are often helpful. Diagrams would be impossible to understand when working with more than about four or five sets.

Example 3

For any integer n, the *ideal* n**Z** is defined to be the set,

$$\{x \in \mathbf{Z} \mid x = ny \text{ for some } y \in \mathbf{Z}\}$$

(i.e. the set of all multiples of n). Prove that $n\mathbf{Z} \subset m\mathbf{Z}$ if and only if n is a multiple of m.

Proof

The proof will look like this:

Let $n\mathbf{Z} \subset m\mathbf{Z}$.

Therefore, n is a multiple of m.

Let n be a multiple of m.

Therefore, n**Z** ⊂ m**Z**.

Inside the second half, there must be,

Let, x ∈ n**Z**.

So, x ∈ m**Z**,

in order to derive, n**Z** ⊂ m**Z**.

The full proof is:

Let,

$$n\mathbf{Z} \subset m\mathbf{Z}$$

Now,

$$n \in n\mathbf{Z}$$

So,

$$n \in m\mathbf{Z}$$

Therefore,

$$n = my$$

for some y ∈ **Z** (from the definition of m**Z**).

That is, n is a multiple of m.

So we have proved that if n**Z** ⊂ m**Z** then n is a multiple of m.

Now suppose n is a multiple of m,

So,

$$n = mz$$

for some z ∈ **Z**.

Let,

$$x \in n\mathbf{Z}$$

So,

$$x = ny$$

for some y ∈ m**Z**

$$= mzy$$

since n = mz

That is, x is a multiple of m.

So,

$$x \in m\mathbf{Z}$$

Thus,

$$n\mathbf{Z} \subset m\mathbf{Z}$$

So we have proved that if n is a multiple of m, then $n\mathbf{Z} \subset m\mathbf{Z}$.

Altogether we have proved that $n\mathbf{Z} \subset m\mathbf{Z}$ if and only if n is a multiple of m. ▮

Notes

1. The step:

   $$n\mathbf{Z} \subset m\mathbf{Z}$$

   Now,

   $$n \in n\mathbf{Z}$$

   So,

   $$n \in m\mathbf{Z}$$

   is an example of the more general argument form,

   All As are Bs
   x is an A
   therefore, x is a B.

   That is, what is true of all As is true of any particular one. There is another example of this type of argument in Exercise 14(b) below.
2. The set with no members is called the *empty set*, and is denoted $\varnothing$. (Does the empty set really exist? Philosophically inclined readers may like to regard it as a useful fiction, like the number zero.) The empty set is considered to be a subset of *every* set. This has some technical advantages, such as making the number of subsets of a set of n elements equal to 2^n. (See Example 5 in Chapter 8.) Now,

   $$\varnothing \subset A$$

   is equivalent to the "if" statement "If $x \in \varnothing$ then $x \in A$". However, $x \in \varnothing$ is false (for any x), so we must make a convention about statements "If p then q" in cases where p is false. The convention is to count all such statements as true. This is consistent with counting $\varnothing \subset A$ as true. Similarly,

   $$\varnothing \subset A$$

   is equivalent to the "all" statement,

   "All members of $\varnothing$ are members of A"

   but there are no members of $\varnothing$. The convention is to count "All As are Bs" as true if there are no As.

   These rather odd conventions are technically useful, for example in starting inductions where the case $n = 1$ (or $n = 0$) means an empty set is being considered. The point of the conventions is to separate the question "Does being a B follow from being an A?", from the question "Are there any As?".

Linear algebra

The *span* of a set of vectors is the set of all linear combinations of them. We write this in symbols thus:

If,

$$S = \{\underset{\sim}{v}_1, \ldots, \underset{\sim}{v}_n\}$$

then,

$$\text{span } S = \{\underset{\sim}{v} : \underset{\sim}{v} = a_1\underset{\sim}{v}_1 + \ldots + a_n\underset{\sim}{v}_n \text{ for some scalars } a_1, \ldots, a_n\}$$

For example, in $\mathbf{R}^3$,

$$\text{span } \{(1, 0, 0), (0, 1, 0)\}$$

is the xy-plane, since,

$$\begin{aligned} \text{span } \{(1, 0, 0), (0, 1, 0)\} &= \{v : v = a(1, 0, 0) + b(0, 1, 0)\} \\ &= \{(a, b, 0)\} \end{aligned}$$

that is, the set of all vectors with the z-coordinate zero, which is the xy-plane.

This example gives some sense to the word "span". The two vectors are a kind of "framework" that is enough to "support" the whole plane, in the sense that every point in the plane can be got by extending out from the two vectors (1, 0, 0) and (0, 1, 0). (See Figure 7.5.)

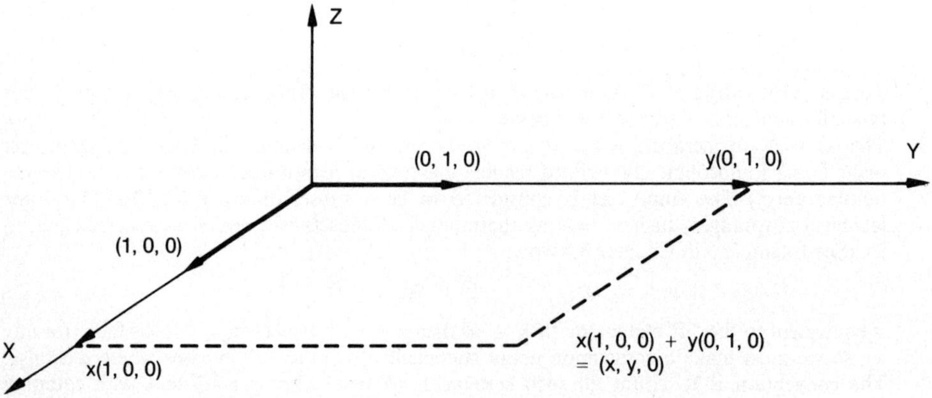

Figure 7.5

Example 4

Prove that the span of (1, 1) and (1, 2) consists of all vectors (a, b), (that is, span $\{(1, 1), (1, 2)\} = \mathbf{R}^2$).

This asks us to show that *every* point (a, b) is of the form $x(1, 1) + y(1, 2)$ for some $x, y \in \mathbf{R}$. This statement has two quantifiers, an "all" and a "some".

Proof

Suppose $(a, b) = x(1, 1) + y(1, 2)$. (We don't know that x, y exist yet; the aim is to show that they do.)

Then,

$$\begin{aligned}(a, b) &= (x, x) + (y, 2y)\\ &= (x + y,\ x + 2y)\end{aligned}$$

So,

$$a = x + y$$

and,

$$b = x + 2y$$

Subtracting the first equation from the second equation,

$$b - a = y$$

and substituting this in the first,

$$a = x + b - a$$

so,

$$2a - b = x$$

This shows that for *all* (a, b), we can find x and y (namely $x = 2a - b$ and $y = b - a$), such that,

$$(a, b) = x(1, 1) + y(1, 2)$$

Therefore every (a, b) is a linear combination of (1, 1) and (1, 2).

That is,

$$\mathbf{R}^2 \subset \text{span}\{(1, 1), (1, 2)\}$$

Also, it is obvious that,

$$\text{span}\{(1, 1), (1, 2)\} \subset \mathbf{R}^2$$

since every linear combination of (1, 1) and (1, 2) is a vector in $\mathbf{R}^2$.

Therefore $\text{span}\{(1, 1), (1, 2)\} = \mathbf{R}^2$. ▮

(We have applied the techniques above: To show two sets are equal, we show that everything in the first is in the second, and everything in the second is in the first.)

Notes

1. Usually the set S is a small finite set, as in Example 4, while span S is an infinite set. However, sense can still be given to the definition if S itself is infinite.
2. The following sentences all mean the same:
   "Span {(1, 0, 0), (0, 1, 0)} is the xy-plane";
   "The span of (1, 0, 0) and (0, 1, 0) is the xy-plane";
   "(1, 0, 0) and (0, 1, 0) span the xy-plane".
   Notice in the first two, "span" is a noun while in the last it is a verb.

The following sentences are meaningless:
"(1, 0, 0) and (0, 1, 0) span";
"S is a span";
"The xy-plane is a span".

Example 5

Prove that span{(2, 4), (1, 2)} $\neq$ $\mathbf{R}^2$.

As in Example 4, it is obvious that, span {(2, 4), (1, 2)} $\subset$ $\mathbf{R}^2$, so we have to show that $\mathbf{R}^2$ is not a subset of span{(2, 4), (1, 2)}. That is, we have to show that span{(2, 4), (1, 2)} is not the whole of $\mathbf{R}^2$.

Proof

This is a "not all" statement: "Not all (a, b) are of the form x(2, 4) + y(1, 2)". To show this we exhibit a particular (a, b) which is not of the form x(2, 4) + y(1, 2). We can see that anything of the form x(2, 4) + y(1, 2) will have the second co-ordinate twice the first, so we choose a vector which does not have that property, say (1, 3). The proof is:

If,

$$(1, 3) = x(2, 4) + y(1, 2)$$

then,

$$\begin{aligned}(1, 3) &= (2x, 4x) + (y, 2y)\\ &= (2x + y,\ 4x + 2y)\end{aligned}$$

so,

$$1 = 2x + y$$

and,

$$\begin{aligned}3 &= 4x + 2y\\ &= 2(2x + y)\\ &= 2.1\\ &= 2\end{aligned}$$

but,

$$3 \neq 2$$

so (1, 3) cannot be of the form x(2, 4) + y(1, 2).

So, (1, 3) is not in the span of (2, 4) and (1, 2).

Therefore, the span of (2, 4) and (1, 2) does not include all points (a, b).

So span {(2, 4), (1, 2)} $\neq$ $\mathbf{R}^2$. ▌

Calculus

The *general solution* of a differential equation is the set of all solutions. Thus, when we say, "The general solution of $\frac{df}{dx} = f$ is Ae^x", we mean that the set of all solutions of $df/dx = f$ is,

$$\{f : f = Ae^x \text{ for some real number } A\}$$

Example 6

If S is the set of all solutions of,

$$\frac{df}{dx} = f$$

and T is the set of all solutions of,

$$\frac{d^2f}{dx^2} = f$$

show that,

$$S \subset T$$

Method 1

Solving the differential equations, we find,

$$S = \{f : f = Ae^x \text{ for some real number } A\}$$

and,

$$T = \{f : f = Ae^x + Be^{-x} \text{ for some real numbers } A, B\}$$

and clearly,

$$S \subset T$$

Method 2

Let,

$$f \in S$$

So,

$$\frac{df}{dx} = f$$

Differentiating,

$$\frac{d^2f}{dx^2} = \frac{df}{dx}$$

but this is f, from the line above.

So f satisfies,

$$\frac{d^2f}{dx^2} = f$$

That is,

$$f \in T$$

Which method is better? See Exercise 23.

Exercises

(Gradings of exercises: * easy, ** moderate, *** difficult.)

***1.** (a) Write down an "all" statement equivalent to, "$A \subset B$". Then write down your "all" statement in "if" form.
(b) Use this "if" statement to explain the form of the proof of the statement, "$A \subset B$".

***2.** The truth of the relation, $B \subset (A \cup B)$, may be verified from a Venn diagram, but this does not constitute a proof. How would you prove this statement?

***3.** Prove that, "If $A \subset B$ and $B \subset C$ then $A \subset C$" (without using a Venn diagram).

***4.** (a) For any two sets A and B explain what is meant by:
(i) $A \cap B$
(ii) $A \cup B$
(b) Prove that, $(A \cap B) \subset B$

***5.** (a) Write down an "if and only if" statement equivalent to, $A = B$.
(b) Use this to explain the form of the proof of the statement, $A = B$.

***6.** (a) How would you prove the statement, $B \cup B = B$?
(b) Prove it.

***7.** Prove that $(A \cap B) \subset (A \cup B)$

***8.** (a) Prove that $B \cup \varnothing = B$.
(b) Simplify the following:
(i) $B \cup B$
(ii) $B \cap B$
(iii) $B \cap \overline{B}$
(iv) $B \cup \varnothing$
(v) $B \cap \varnothing$
(Check that you can prove your answers.)

***9.** (a) Explain by the use of an appropriate Venn diagram that: If $A \subset B$ then $\overline{B} \subset \overline{A}$.

(b) Are the two statements, "All As are Bs", and its contrapositive, "All not-Bs are not-As" logically equivalent?

****10.** Prove that, $A \cup (B \cup C) = (A \cup B) \cup C$.

****11.** (a) Draw appropriate diagrams to verify the truth of the statement, $A \cup (B \cap C) = (A \cup B) \cap (A \cup C)$, then prove it.
(b) Prove that, $A \cap (B \cup C) = (A \cap B) \cup (A \cap C)$.

****12.** De Morgan's laws state:

$$\overline{(A \cup B)} = \bar{A} \cap \bar{B}$$
$$\overline{(A \cap B)} = \bar{A} \cup \bar{B}$$

(a) Draw appropriate Venn diagrams to verify the truth of, $\overline{(A \cup B)} = \bar{A} \cap \bar{B}$.
(b) Prove, $\overline{(A \cup B)} = \bar{A} \cap \bar{B}$.
(c) Prove, $\overline{(A \cap B)} = \bar{A} \cup \bar{B}$.

****13.** Comment on the reasoning,

Most As are Bs
Most As are Cs

Therefore some Bs are Cs.

****14.** A set of real numbers is called *bounded* if it does not "go to infinity". More precisely, S is bounded if there exist real numbers M, N such that for all $s \in S$, $M < s < N$. (For example, the set of real numbers x such that $1 < x^3 < 2$ is bounded, since for all $x \in S$, $1 < x < 1.5$.)
(a) Give an example of a set which is not bounded.
(b) Prove that if S is bounded and $T \subset S$, then T is bounded.
(c) Prove that any finite set of real numbers is bounded.
(d) Prove that if T is not bounded and $T \subset S$, then S is not bounded.
(e) Prove that if S and T are bounded, then $S \cap T$ is bounded.
(f) If S and T are bounded, is $S \cup T$ always bounded? Prove your answer.
(g) Let S and T be bounded. Let,

$$U = \{u \in \mathbf{R}: u = s + t \text{ for some } s \in S, t \in T\}$$

Show that U is bounded. It might help to calculate some examples first, say,

$$S = [0, 1], T = [2, 3]$$

(The set U is sometimes denoted $S + T$, since it is the set of all sums of something in S with something in T.)

****15.** A region in the plane is called *convex* if the line segment joining any two points in the region lies wholly insido the region. For example, an ellipse, a parallelogram, a triangle and a straight line are convex, but an annulus and a star-shaped region are not. In symbols, R is convex if, for all (x_1, y_1) and (x_2, y_2) in R, $(\lambda x_1 + (1 - \lambda)x_2, \lambda y_1 + (1 - \lambda)y_2) \in R$ for all $\lambda \in [0, 1]$.
(a) Prove that if R and S are convex, then $R \cap S$ is convex.

(b) If R and S are convex, is $R \cup S$ always convex? Prove your answer.
(c) Prove that if R is convex, then the reflection of R in the x-axis is convex.
(d) If R is convex, is the set,

$$2R = \{(x, y) : (x, y) = (2x', 2y') \text{ for some } (x', y') \in R\}$$

always convex? Prove your answer and illustrate with a diagram.

****16.** Prove that $n\mathbf{Z} = m\mathbf{Z}$ if and only if $n = m$ or $n = -m$

*****17.** Fill in the blank and prove the theorem:

"$A \cap (B \cup C) = (A \cap B) \cup C$ if and only if ——"

Linear algebra

***18.** (a) Show that $\mathbf{R}^2 = \text{span}\{(1, 1), (1, -1)\}$
(b) Show that $\mathbf{R}^2 \neq \text{span}\{(1, 1), (-3, -3)\}$

***19.** Show that (1, 2, 3) is not in the span of (1, 1, 1) and (1, 0, 2)

***20.** Show that any vector (a, b, c) such that $a + b + c = 0$ is in the span of $\{(1, 1, -2), (1, 0, -1)\}$

****21.** Show that the span of any set of vectors $\underset{\sim}{v}_1, \ldots, \underset{\sim}{v}_n$ in a vector space V is a subspace of V (that is, is closed under addition and scalar multiplication).

*****22.** Show that, $\text{span}\{(a, b), (c, d)\} = \mathbf{R}^2$ if and only if $ad - bc \neq 0$.

Calculus

***23.** Let S be the set of all solutions of,

$$\frac{d^2f}{dx^2} = f$$

and let T be the set of all solutions of,

$$\frac{d^4f}{dx^4} = f$$

Show that $S \subset T$.

***24.** Show that the set of solutions of,

$$\frac{df}{dx} = 2x$$

is infinite.

****25.** The general solution of,

$$\frac{d^2y}{dx^2} = y$$

may be written as $Ae^x + Be^{-x}$.

(a) Check that $\sinh x$ and $\cosh x$ are solutions of this differential equation.
(b) Prove that $y = C \sinh x + D \cosh x$ is a solution, for any real numbers C and D.
(c) Is,

$$\{(y : y = C \sinh x + D \cosh x \text{ for some } C, D \in \mathbf{R}\}$$

equal to the set,

$$\{y : y = Ae^{x} + Be^{-x} \text{ for some } A, B \in \mathbf{R}\}?$$

8

Proof by mathematical induction

Mathematical induction is a special method of proof used to prove statements about all the natural numbers. For example,

"$n^3 - n$ is always divisible by 3"

"The sum of the first n integers is $\frac{n(n+1)}{2}$"

The first of these makes a different statement for each natural number n. It says, $1^3 - 1$, $2^3 - 2$, $3^3 - 3$, and so on, are all divisible by 3. The second also makes a statement for each n. It says,

for n = 1,

$$1 = \frac{1.2}{2}$$

for n = 2,

$$1 + 2 = \frac{2.3}{2}$$

for n = 100,

$$1 + 2 + 3 + \ldots + 100 = \frac{100.101}{2}$$

and in general, for any natural number n,

$$1 + 2 + 3 + \ldots + n = \frac{n(n+1)}{2}$$

The idea behind proof by mathematical induction is a simple one: The natural numbers form a list, so that if we start at the beginning and keep going from one number to the next, we eventually get to any given number. Thus:

$$1 \to 2 \to 3 \to 4 \ \ldots \ 100 \to 101 \ \ldots \ n \to n+1 \ \ldots$$

Therefore, if we can prove that some statement involving n is true for $n = 1$ (the beginning of the list) and that the truth of the statement for n implies its truth for $n + 1$ (so we can get from any number to the next), then we have proved the statement for all n.

Example 1

Prove by mathematical induction that $n^3 - n$ is divisible by 3 for all natural numbers n.

Proof

For $n = 1$,

$$\begin{aligned} n^3 - n &= 1 - 1 \\ &= 0 \end{aligned}$$

which is divisible by 3.

Assume the statement is true for *some* number n, that is, $n^3 - n$ is divisible by 3. Now,

$$\begin{aligned} (n + 1)^3 - (n + 1) &= n^3 + 3n^2 + 3n + 1 - n - 1 \\ &= (n^3 - n) + 3(n^2 + n) \end{aligned}$$

which is $n^3 - n$ plus a multiple of 3.

Since we assumed that $n^3 - n$ was a multiple of 3, it follows that $(n + 1)^3 - (n + 1)$ is also a multiple of 3.

So, since the statement "$n^3 - n$ is divisible by 3" is true for $n = 1$, and its truth for n implies its truth for $n + 1$, the statement is true for all whole numbers n.

Example 2

Prove that,

$$1 + 2 + 3 + \ldots + n = \frac{n(n + 1)}{2}$$

Proof

For $n = 1$,

$$1 = \frac{1.2}{2}$$

This is true.

Assume the truth of the statement for some n, that is,

$$1 + 2 + 3 + \ldots + n = \frac{n(n + 1)}{2}$$

Now,

$$1 + 2 + 3 + \ldots + n + (n + 1) = \frac{n(n + 1)}{2} + (n + 1)$$

by induction assumption,

$$= (n + 1)\left(\frac{n}{2} + 1\right)$$

$$= \frac{(n + 1)(n + 2)}{2}$$

$$= \frac{(n + 1)((n + 1) + 1)}{2}$$

which is the statement with n replaced by n + 1.

So the statement is true for n = 1, and its truth for n implies its truth for n + 1.

Therefore, it is true for all n.

Notes

1. The essential steps of a proof by mathematical induction are: the proof for n = 1; and the proof of the "*if*" proposition, "If the statement is true for n it is true for n + 1." As discussed earlier, the way to prove an "if" proposition is to assume the first part and deduce the second part from it.
2. It is not necessary to start at n = 1. The results proved in Examples 1 and 2 both actually make sense, and are true, for n = 0, and the proofs could have started by showing their truth for n = 0 instead of n = 1. On the other hand, some proofs may need to start later than n = 1. For example, $2^n > n^2$ is true for $n \geqslant 5$, (but is not true for n = 2, 3 or 4). Its proof would start by showing it is true for n = 5, then showing that its truth for n implied its truth for n + 1. This would show the statement was true for all $n \geqslant 5$.
3. Sometimes, to get the truth of the statement for n + 1, it is necessary to assume its truth not just for n, but for n and n − 1, or sometimes for all numbers up to n.

Example 3

The formula for the nth term a_n of the Fibonacci sequence,

1, 1, 2, 3, 5, 8, 13, 21, 34, . . .

is given by,

$$a_n = \begin{cases} 1 \text{ for } n = 1 \text{ and } 2 \\ a_{n-2} + a_{n-1} \text{ for } n > 2 \end{cases}$$

Prove by mathematical induction that,

$$a_n = \frac{(1 + \sqrt{5})^n - (1 - \sqrt{5})^n}{\sqrt{5}\, 2^n}$$

Proof

For n = 1,

$$1 = \frac{(1 + \sqrt{5}) - (1 - \sqrt{5})}{\sqrt{5}\,2}$$

This is true.

For n = 2,

$$1 = \frac{(1 + \sqrt{5})^2 - (1 - \sqrt{5})^2}{\sqrt{5}\,2^2}$$

This is also true.

Assume the truth of the statement for some n − 1 and n, that is,

$$a_{n-1} = \frac{(1 + \sqrt{5})^{n-1} - (1 - \sqrt{5})^{n-1}}{\sqrt{5}\,2^{n-1}}$$

and,

$$a_n = \frac{(1 + \sqrt{5})^n - (1 - \sqrt{5})^n}{\sqrt{5}\,2^n}$$

Now,

$$\begin{aligned}
a_{n+1} &= a_{n-1} + a_n \\
&= \frac{(1 + \sqrt{5})^{n-1} - (1 + \sqrt{5})^{n-1}}{\sqrt{5}\,2^{n-1}} + \frac{(1 + \sqrt{5})^n - (1 + \sqrt{5})^n}{\sqrt{5}\,2^n} \\
&= \frac{4(1 + \sqrt{5})^{n-1} - 4(1 - \sqrt{5})^{n-1} + 2(1 + \sqrt{5})^n - 2(1 - \sqrt{5})^n}{\sqrt{5}\,2^{n+1}} \\
&= \frac{(1 + \sqrt{5})^{n-1}(4 + 2(1 + \sqrt{5})) - (1 - \sqrt{5})^{n-1}(4 + 2(1 - \sqrt{5}))}{\sqrt{5}\,2^{n+1}} \\
&= \frac{(1 + \sqrt{5})^{n-1}(6 + 2\sqrt{5}) - (1 - \sqrt{5})^{n-1}(6 - 2\sqrt{5})}{\sqrt{5}\,2^{n+1}} \\
&= \frac{(1 + \sqrt{5})^{n-1}(1 + \sqrt{5})^2 - (1 - \sqrt{5})^{n-1}(1 - \sqrt{5})^2}{\sqrt{5}\,2^{n+1}} \\
&= \frac{(1 + \sqrt{5})^{n+1} - (1 - \sqrt{5})^{n+1}}{\sqrt{5}\,2^{n+1}}
\end{aligned}$$

which is the statement for n + 1.

So the statement is true for n = 1 and n = 2, and its truth for n − 1 and n implies its truth for n + 1.

Therefore it is true for all n.

Example 4

Prove that every positive integer greater than 1 can be written as a product of primes.

The result was assumed "sufficiently well known" in the proof for Example 2 in Chapter 5 (i.e. that there are infinitely many primes). Now it can be proved properly. The words, "can be written" are simply a metaphor for an existential statement. In this example the statement is:

"For any integer $n > 1$ there exist primes $p_1, \ldots, p_m$ such that $n = p_1 \ldots p_m$."

Of course, some of the primes could be equal.

Proof

If n is a prime, then it is already a product of primes.

If n is not a prime, then $n = n_1 n_2$ where neither n_1 or n_2 are 1.

So, n_1 and n_2 are both less than n.

We take as our induction hypothesis the statement, "Any number less than n (but greater than 1) can be written as a product of primes".

So, n_1 and n_2 are products of primes and therefore $n = n_1 n_2$ is also a product of primes.

So the truth of the result for all numbers less than n implies the truth of the result for n.

The result is true for $n = 2$, so, by induction, it is true for all $n > 1$.

Fermat's Infinite Descent

The original version of proof by mathematical induction was Pierre Fermat's method of "Infinite Descent" (devised around 1650). In some ways it is a more natural way to regard the method. It is illustrated in Example 5, as follows.

Example 5

A set of n elements has 2^n subsets.

(When counting subsets, it must be remembered that, by convention, the empty set and the whole set are regarded as subsets of any set.)

Proof

Denote the number of subsets of a set of n elements by S(n).

Take a set of n elements.

Take one of the elements, say a, and consider separately the subsets which contain a and those which do not.

There are S(n − 1) subsets not containing a (because they are the subsets of the (n − 1) element set consisting of the original set without a).

The number of subsets which do contain a is also S(n − 1), since a subset with a is formed by adding a to one of the subsets without a.

So,

$$\begin{aligned} S(n) &= S(n-1) + S(n-1) \\ &= 2\,S(n-1) \end{aligned}$$

Similarly,

$$S(n-1) = 2\,S(n-2)$$

and so on, so that,

$$S(n) = 2\,S(n-1) = 4\,S(n-2) = \ldots = 2^{n-1}\,S(1)$$

and,

$$S(1) = 2$$

(since a one-element set has two subsets, namely the whole set and the empty set).

Thus,

$$S(n) = 2^n \quad \blacksquare$$

Notes

1. It would be easy to recast this solution in the usual form by beginning, "Assume $S(n) = 2^n$", but it seems more natural to proceed as above, reducing the problem for n to the problem for n − 1, and then "descending". A different proof of this result is given in Chapter 11.
2. Fermat's statement of his method of infinite descent also included a use of proof by contradiction. To prove a statement about n, he suggested showing that if the statement was *false* for n, it would also be false for some number smaller than n. Then, by "descending", one would find the statement to be false for n = 1. Checking would show that the result was in fact true for n = 1, so it could be concluded that the result was true for any n. Later writers have preferred not to use proof by contradiction here, since it is not necessary. (C. B. Boyer, *A History of Mathematics* (1968), Wiley, New York, p. 387.)

If there is a choice between proving a statement by mathematical induction and proving it by some other method, the other method is often preferred because it gives more insight. For example, the statement in Example 1. in this Chapter, can be better proved thus:

Prove $n^3 - n$ is a multiple of 3.

Proof

$$n^3 - n = n(n^2 - 1) = (n-1)n(n+1)$$

which is the product of 3 consecutive integers.

One of these must be a multiple of 3, so, $n^3 - n$ is a multiple of 3. $\blacksquare$

(The fact that one of any three consecutive numbers is a multiple of 3 can be assumed to be sufficiently obvious to serve as the basis of a proof. Nevertheless, if we were to ask for a proof of this fact from more basic facts about numbers, we would find ourselves using induction again. Thus, there may be an induction hidden even in this proof.)

Example 6

Prove,

$$1 + 2 + 3 + \ldots + n = \frac{n(n + 1)}{2}$$

Proof

This proof is suggested by the 2 on the bottom of the formula, which indicates it might be helpful to write down the sum twice:

$$1 + 2 + 3 + \ldots + n = \frac{1}{2}\begin{pmatrix} 1 + & 2 & + & 3 & + \ldots + & (n - 1) & + n + \\ n + & (n - 1) & + & (n - 2) & + \ldots + & 2 & + 1 \end{pmatrix}$$

The second sum has been written backwards and underneath the first so we can compare the numbers written together. Each pair adds up to n + 1:

$$= \frac{1}{2}(\underbrace{(n + 1) + (n + 1) + \ldots + (n + 1)}_{n \text{ times}})$$

$$= \frac{1}{2}(n + 1)n$$

which is the answer.

Notes

1. A great disadvantage of the method of mathematical induction is that the formula to be proved true must be known, or at least conjectured, to be true before we start. This method gives no help at all in finding the formula. However, to some extent this is a problem with any kind of proof. There is a possible way of finding the formula to be proved. Start by working out the first few cases, then try to see a pattern, and from this conjecture what the general result should be. Some examples will be given in Chapter 14.
2. "Mathematical induction" must be distinguished from "induction" in general. See Chapter 14 for further discussion.

Proofs by induction are commonly used in computing applications when establishing facts about the strings of symbols in computer languages.

Example 7

A formal language to be used for simple algebra uses the symbols,

x y z ( ) +

The *words* of the language are strings of symbols formed according to the rules:

1. x, y and z are words.
2. If A and B are words, so is (A)(B).
3. If A and B are words, so is (A) + (B).

Nothing is a word other than the words formed according to the rules, 1., 2. and 3.

For example, ((x)(z)) + (z) is a word, since x and z are words (Rule 1.) and therefore so is (x)(z) (Rule 2.) and hence ((x)(z)) + (z) (Rule 3.). However, (x) + z is not a word.

Prove that any word in this language has the same number of ('s and)'s.

Proof

We use induction on the length of words, that is, the number of symbols in them.

A word of length 1 is either x, y or z so the number of ('s and)'s in a word of length 1 is zero.

So the result is true for words of length 1.

Assume the result is true for words of lengths less than n.

A word of length n is either of the form (A) (B) or of the form (A) + (B), for some words A, B of length less than n.

By induction assumption, A and B have the same number of ('s and)'s.

Therefore, (A) (B) and (A) + (B) also have the same number of ('s and)'s.

So the result is true for words of any length.

Exercises

(Gradings of exercises: * easy, ** moderate, *** difficult.)

***1.** Prove by mathematical induction the following:

(a) $1^2 + 2^2 + 3^2 + \ldots + n^2 = \frac{1}{6}n(n + 1)(2n + 1)$

(b) $1^3 + 2^3 + 3^3 + \ldots + n^3 = \left(\frac{n(n + 1)}{2}\right)^2$

(c) $2 + 4 + 6 + \ldots + 2n = n(n + 1)$

(d) $1 + 3 + 5 + \ldots + (2n - 1) = n^2$

(e) For all positive integers n, $n^2 - n$ is even.

****2.** Show that, $1 - \frac{1}{2} + \frac{1}{3} - \frac{1}{4} + \frac{1}{5} + \ldots + (-1)^{n-1}\frac{1}{n}$ is always positive.

****3.** Prove, by mathematical induction, the formula for the sum of the first n terms of an arithmetic progression:

$$a + (a + d) + (a + 2d) + (a + 3d) + \ldots + (a + (n - 1)d) = \frac{n}{2}(2a + (n - 1)d)$$

****4.** Prove, by mathematical induction, the formula for the sum of the first n terms of a geometric progression:

$$a + ar + ar^2 + \ldots + ar^{n-1} = \frac{a(1 - r^n)}{1 - r}, \; r \neq 1$$

****5.** (a) Prove, by mathematical induction, that if n is a whole number then,

$$n^3 + 3n^2 + 2n$$

is divisible by 6.

(b) Prove the same result without mathematical induction by first factorising $n^3 + 3n^2 + 2n$.

****6.** Show that a number which consists of 3^n equal digits is divisible by 3^n.

****7.** (a) Calculate,

$$\frac{1}{2!} + \frac{2}{3!} + \frac{3}{4!} + \ldots + \frac{n}{(n + 1)!}$$

for a few small values of n.

(b) Make a conjecture about a formula for this expression.

(c) Prove your conjecture by mathematical induction.

****8.** The following is a famous fallacy that uses the method of mathematical induction. Explain what is wrong with it.

"Theorem"

Everything is the same colour.

"Proof"

Let P(n) be the statement, "In every set of n things, all the things have the same colour".
We will show that P(n) is true for all n = 1, 2, 3, . . . , so that *every* set consists of things of the same colour.
Now, P(1) is true, since in every set with only one thing in it, everything is obviously of the same colour.
Now, suppose P(n) is true.
Consider any set of n + 1 things.
Take an element of the set, a. The n things other than a form a set of n things, so they are all the same colour (since P(n) is true).
Now take a set of n things out of the n + 1 which does include a.
These are also all the same colour, so a is the same colour as the rest.
Therefore P(n + 1) is true.

****9.** If a sequence a_n satisfies,

$$a_{n+1} = \frac{a_n}{a_n + 1}$$

Show that,

$$a_n = \frac{a_0}{na_0 + 1}$$

***10. Suppose that we draw on a plane n lines in "general position" (i.e. with no three concurrent, and no two parallel). Let s_n be the number of regions into which these lines divide the plane, for example, $s_3 = 7$ in Figure 8.1.

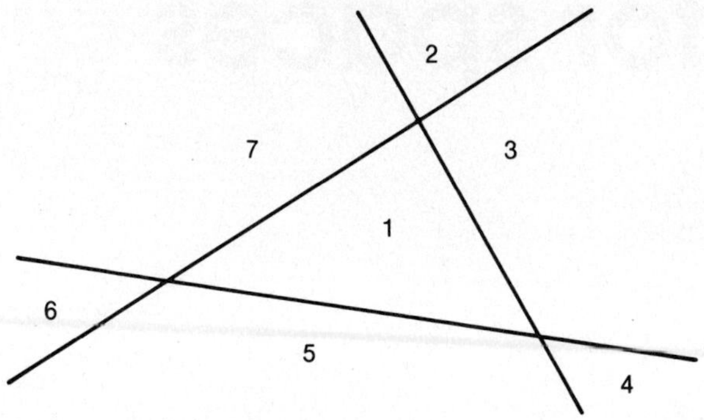

Figure 8.1

(a) By drawing diagrams, find s_1, s_2, s_3, s_4 and s_5
(b) From these results, make a conjecture about a formula for s_n
(c) Prove this formula by mathematical induction.

***11. The "pigeonhole principle" states: If n things are put in fewer than n holes, some hole has at least two things. Prove this.

***12. "Fermat's Small Theorem" states that $n^p - n$ is divisible by p for any prime p. Prove this.

***13. If a set has a binary operation * satisfying the associative law:

$$(a * b) * c = a * (b * c)$$

show that any product of n elements is independent of the order of brackets.

Calculus

**14. Show that the nth derivative of x^n is n!.

**15. Use integration by parts to show that for $n \geqslant 2$,

$$\int_0^{\pi} \sin^n x \, dx = \frac{n-1}{n} \int_0^{\pi} \sin^{n-2} x \, dx$$

and hence find,

$$\int_0^{\pi} \sin^8 x \, dx$$

9
Vector spaces

In Chapter 2 the importance of generalisation was discussed. Generalisation allows all the individual facts to be proved at the same time, and provides a clearer understanding by showing what aspects of the situation are relevant to the result. For example, instead of dealing with the separate facts,

$$\sqrt[2]{2!} < \sqrt[3]{3!},\ \sqrt[3]{3!} < \sqrt[4]{4!}$$

and so on, it is preferable to prove the "all" statement,

$$\text{For all whole numbers n, } \sqrt[n]{n!} < \sqrt[n+1]{(n+1)!}$$

In *abstract algebra*, which includes the general theory of *vector spaces*, the generalisation process continues. This is because it is recognised that, fundamentally, the same thing is happening in what at first appear to be different contexts. For example, in $\mathbf{R}^3$,

$$(1, 2, 4) + (3, -1, 7) = (4, 1, 11)$$

while with polynomials,

$$(x^2 + 2x + 4) + (3x^2 - x + 7) = 4x^2 + x + 11$$

and with forces,

$$(\underset{\sim}{i} + 2\underset{\sim}{j} + 4\underset{\sim}{k}) + (3\underset{\sim}{i} - \underset{\sim}{j} + 7\underset{\sim}{k}) = 4\underset{\sim}{i} + \underset{\sim}{j} + 11\underset{\sim}{k}$$

(where $\underset{\sim}{i} + 2\underset{\sim}{j} + 4\underset{\sim}{k}$ indicates a force with components 1 newton in the x direction, 2 in the y direction and 4 in the z direction, and so on).

It is clear that under the different notations, the same thing is happening mathematically. The theory of vector spaces provides a setting in which such cases can be considered together. In fact, some of the results of abstract vector space theory have been implicitly considered in earlier chapters. Consider Example 7 in Chapter 6: "Show that if $\underset{\sim}{u}$ and $\underset{\sim}{v}$ are linearly independent vectors, then $\underset{\sim}{u} + \underset{\sim}{v}$ and $\underset{\sim}{u} - \underset{\sim}{v}$ are also independent". This result applies if $\underset{\sim}{u}$ and $\underset{\sim}{v}$ are vectors in $\mathbf{R}^3$, but the proof does not depend on any special facts about $\mathbf{R}^3$. It applies equally well if $\underset{\sim}{u}$ and $\underset{\sim}{v}$ are polynomials, or forces, or matrices, or functions, or any other things which it makes sense to add, subtract and multiply by scalars (recall that the notion of linear independence involves multiplication by scalars as well as addition).

Consider a set of all vectors of the same kind, for example the collection of all the velocities that a body can have. We can add any two of these velocities to obtain another velocity (the sum) which is again in that set, that is, it is another velocity that the body could have. We can also multiply any velocity in the set by any scalar to obtain another velocity in the set. So the set of all vectors of the same kind is said to be a "vector space".

There are some other kinds of things which, though they do not have magnitude and direction, can be added together and multiplied by scalars in the same way as velocities and forces. For example, any two 2×3 matrices can be added together to give another 2×3 matrix. Thus,

$$\begin{pmatrix} 1 & 2 & 1 \\ 3 & 1 & 0 \end{pmatrix} + \begin{pmatrix} 1 & 1 & 1 \\ 1 & 1 & 1 \end{pmatrix} = \begin{pmatrix} 2 & 3 & 2 \\ 4 & 2 & 1 \end{pmatrix}$$

Similarly, we can multiply any 2×3 matrix by a number, and the answer is another 2×3 matrix:

$$3\begin{pmatrix} 1 & 2 & 1 \\ 3 & 1 & 0 \end{pmatrix} = \begin{pmatrix} 3 & 6 & 3 \\ 9 & 3 & 0 \end{pmatrix}$$

We therefore say that the set of all 2×3 matrices (denoted M_{23}) is a *vector space*, and we call the members of it, the individual 2×3 matrices, *vectors*.

From the above instances, it can be seen that we apply the name *vector space* to any system of things which can be added together, and multiplied by numbers. The things in any such system are then called *vectors*. See Table 9.1 which lists examples of vector spaces.

The elements in each of the systems given as examples of vector spaces can be added together, and can be multiplied by numbers. This is the common feature of all vector spaces. Notice that in each case, vector + vector = vector (in the same vector space) and scalar $\times$ vector = vector. Two vectors from different vector spaces cannot be added together (except in the case where one vector space is actually part of another, as in examples 8 and 7 in Table 9.1).

There are eight *axioms* for a vector space which make precise what is to count as "adding" and "multiplying by scalars" (if these axioms were not satisfied we would not call the operations "addition" and "scalar multiplication").

A vector space V is any set with:

(a) an operation + such that for any vectors $\underset{\sim}{u}, \underset{\sim}{v} \in V$, $\underset{\sim}{u} + \underset{\sim}{v} \in V$; and
(b) multiplication of vectors by scalars such that for $\alpha \in \mathbf{R}$ and $\underset{\sim}{u} \in V$, $\alpha \underset{\sim}{u} \in V$

which satisfy the axioms–

1. $\underset{\sim}{u} + \underset{\sim}{v} = \underset{\sim}{v} + \underset{\sim}{u}$ (commutativity)
2. $(\underset{\sim}{u} + \underset{\sim}{v}) + \underset{\sim}{w} = \underset{\sim}{u} + (\underset{\sim}{v} + \underset{\sim}{w})$ (associativity)
3. There exists a vector $\underset{\sim}{0} \in V$ (called the zero vector) such that for all $\underset{\sim}{u} \in V$ $\underset{\sim}{u} + \underset{\sim}{0} = \underset{\sim}{u}$
4. For all $\underset{\sim}{u} \in V$ there exists a vector $-\underset{\sim}{u}$ in V such that $\underset{\sim}{u} + (-\underset{\sim}{u}) = 0$
5. $\alpha(\underset{\sim}{u} + \underset{\sim}{v}) = \alpha\underset{\sim}{u} + \alpha\underset{\sim}{v}$ } distributivity
6. $(\alpha + \beta)\underset{\sim}{u} - \alpha\underset{\sim}{u} + \beta\underset{\sim}{u}$ }
7. $(\alpha\beta)\underset{\sim}{u} = \alpha(\beta\underset{\sim}{u})$
8. $1\underset{\sim}{u} = \underset{\sim}{u}$

Table 9.1 Examples and non-examples of vector spaces

A. EXAMPLES OF VECTOR SPACES

Vector space	*Some typical vectors*
1. The set of all possible forces on a body.	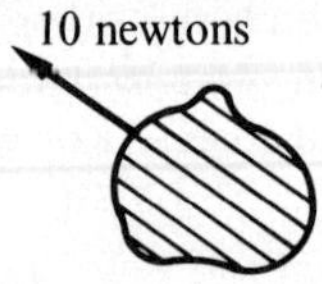
2. The set of all pairs of real numbers (this vector space is denoted $\mathbf{R}^2$).	(1, 3), (−20, 1.73)
3. The set of all triples of real numbers, denoted $\mathbf{R}^3$ (this space can be pictured as the arrows coming out from the origin in 3 dimensions).	(4, −3.1, 5.28)
4. For any positive integer n, the set of n-tuples of real numbers, or lists of n real numbers $(a_1, \ldots, a_n)$, denoted $\mathbf{R}^n$. (This gives, in fact, infinitely many different vector spaces, one for each n.)	(4, 2.7, . . . , 25)
5. The set of 2×3 matrices, denoted M_{23}.	$\begin{pmatrix} 1 & 0.3 & 0 \\ -7 & 1.1 & 2.57 \end{pmatrix}$
More generally:	
6. For any natural numbers m and n, the set of all $m \times n$ matrices, denoted M_{mn}.	$\begin{pmatrix} 2 & 3 & -1 & \ldots & 2.1 \\ -5 & .4 & 4.7 & & 9 \\ \vdots & & & & \vdots \\ 8 & \ldots & & & 3.6 \end{pmatrix}$
7. The set of all polynomials.	$1 - x + x^2$, $-\pi x + 2x^{107}$
8. The set of all polynomials of degree at most 3.	$1 - x + x^2 + 7x^3$, $2 - x^3$, $1 + x$

B. NON-EXAMPLES OF VECTOR SPACES

- The set of all people is not a vector space. You cannot multiply a person by a number; or add people.
- The set of integers is not a vector space. They can be added, but multiplying an integer by a real number does not always give an integer.
- The set of polynomials of degree 3 (exactly) is not a vector space, since adding two cubics does not always give a cubic. For example $(1 + 2x^2 + x^3) + (3 - x^3)$ is a quadratic.

(Axioms 1–2 and 5–8 are understood to be "all" statements. Thus 7 asserts that for *all* $\underset{\sim}{u} \in V$ and all scalars $\alpha, \beta \in \mathbf{R}$, $(\alpha\beta)\underset{\sim}{u} = \alpha(\beta\underset{\sim}{u})$.)

So the axioms are all universal statements except 3 and 4 which have both a "some" and an "all".

Let us check that some of these axioms are true for $\mathbf{R}^3$:

Let,

$$\underset{\sim}{a} = (a_1, a_2, a_3)$$

and,

$$\underset{\sim}{b} = (b_1, b_2, b_3)$$

be two elements of $\mathbf{R}^3$ and $\alpha \in \mathbf{R}$.

Then,

$$\underset{\sim}{a} + \underset{\sim}{b} = (a_1 + b_1, a_2 + b_2, a_3 + b_3)$$

which is in $\mathbf{R}^3$,

and,

$$\alpha(\underset{\sim}{a}) = (\alpha a_1, \alpha a_2, \alpha a_3)$$

which is in $\mathbf{R}^3$.

To show that $\mathbf{R}^3$ is a vector space, we would have to show that each of the eight axioms holds. Here we will only show that Axiom 3 holds; try some of the others as an exercise.

$$\underset{\sim}{0} = (0, 0, 0)$$

is the zero vector in $\mathbf{R}^3$ because for any (a_1, a_2, a_3),

$$\begin{aligned}(a_1, a_2, a_3) + (0, 0, 0) &= (a_1 + 0,\ a_2 + 0,\ a_3 + 0)\\ &= (a_1, a_2, a_3)\end{aligned}$$

So there does exist a vector $\underset{\sim}{0} \in \mathbf{R}^3$ such that,

$$\underset{\sim}{u} + \underset{\sim}{0} = \underset{\sim}{u}$$

for all $\underset{\sim}{u} \in \mathbf{R}^3$.

Since the axioms are true of any vector space, any statement that follows from the axioms will also be true of any vector space.

Here are two examples of theorems that can be derived from the axioms.

Theorem

There is only one zero vector (i.e. only one vector $\underset{\sim}{0}$ such that for all $\underset{\sim}{u} \in V$, $\underset{\sim}{u} + \underset{\sim}{0} = \underset{\sim}{u}$).

Proof

Suppose there were two zero vectors $\underset{\sim}{0}$ and $\underset{\sim}{0}'$ (we aim to show that $\underset{\sim}{0} = \underset{\sim}{0}'$).

Since $\underset{\sim}{0}$ is a zero vector,

$$\underset{\sim}{u} + \underset{\sim}{0} = \underset{\sim}{u}$$

for all

$$\underset{\sim}{u} \in V$$

This is true in particular if we take $\underset{\sim}{u} = \underset{\sim}{0}'$,

So,

$$\underset{\sim}{0}' + \underset{\sim}{0} = \underset{\sim}{0}'$$

But,

$$\underset{\sim}{0}' + \underset{\sim}{0} = \underset{\sim}{0} + \underset{\sim}{0}'$$

by commutativity of addition (Axiom 1)
and since $\underset{\sim}{0}'$ is a zero vector,

$$\underset{\sim}{0} + \underset{\sim}{0}' = \underset{\sim}{0}$$

Therefore, $\underset{\sim}{0}' = \underset{\sim}{0}$ as we had to show ▮

Theorem

For all $\underset{\sim}{u} \in V$, $0\underset{\sim}{u} = \underset{\sim}{0}$
(i.e. multiplying any vector by the scalar 0 gives the zero vector defined by Axiom 3.)

Proof

Since,

$$\underset{\sim}{u} = 1\underset{\sim}{u}$$

(by Axiom 8)

$$\begin{aligned} 0\underset{\sim}{u} + \underset{\sim}{u} &= 0\underset{\sim}{u} + 1\underset{\sim}{u} \\ &= (0 + 1)\underset{\sim}{u} \end{aligned}$$

(by Axiom 6)

$$\begin{aligned} &= 1\underset{\sim}{u} \\ &= \underset{\sim}{u} \end{aligned}$$

(by Axiom 8).

So,

$$(0\underset{\sim}{u} + \underset{\sim}{u}) + (-\underset{\sim}{u}) = \underset{\sim}{u} + (-\underset{\sim}{u})$$

(where $-\underset{\sim}{u}$ is the vector defined in Axiom 4)
but the left hand side is,

$$0\underset{\sim}{u} + (\underset{\sim}{u} + (-\underset{\sim}{u}))$$

(by Axiom 2)

while the right hand side is $\underset{\sim}{0}$ (by Axiom 4).

So we have,

$$0\underset{\sim}{u} + (\underset{\sim}{u} + (-\underset{\sim}{u})) = \underset{\sim}{0}$$

$$0\underset{\sim}{u} + \underset{\sim}{0} = \underset{\sim}{0}$$

(by Axiom 4)

$$0\underset{\sim}{u} = \underset{\sim}{0}$$

(by Axiom 3). ▌

Notes

1. The art of deriving apparently obvious results from axioms is more difficult that it looks. There is not much intuition to go on, and one must treat the problem as one would a game like chess—one must look for some sequence of allowed moves that will achieve the desired result. In the second case above, the essential idea used in constructing the proof of, "$0\underset{\sim}{u} = \underset{\sim}{0}$" was to say, "We must show that $0\underset{\sim}{u}$ satisfies the definition of the zero vector $\underset{\sim}{0}$, that is, adding $0\underset{\sim}{u}$ to another vector must not change that vector. What vector could we add to $0\underset{\sim}{u}$, that might allow us to make progress? Perhaps $0\underset{\sim}{u} + \underset{\sim}{u}$ . . .". The rest of the proof is then not too difficult; progress is guided by the idea of getting to the zero vector, as required by the result.
2. Some steps were not justified in terms of any axioms. They were:

   (a)

   $$(0 + 1)\underset{\sim}{u} = 1\underset{\sim}{u}$$

   (which is true simply because we have added the numbers 0 and 1 to give 1—we may take for granted the usual truths of arithmetic); and

   (b) the two steps obtained by adding a vector to both sides of an equation:

   from,

   $$\underset{\sim}{u} = 1\underset{\sim}{u}$$

   we concluded,

   $$0\underset{\sim}{u} + \underset{\sim}{u} = 0\underset{\sim}{u} + 1\underset{\sim}{u}$$

   and from,

   $$(0\underset{\sim}{u} + \underset{\sim}{u}) = \underset{\sim}{u}$$

   we concluded,

   $$(0\underset{\sim}{u} + \underset{\sim}{u}) + (-\underset{\sim}{u}) = \underset{\sim}{u} + (-\underset{\sim}{u})$$

   The justification here is the general logical property of equality that if $x = y$ then $Fx = Fy$ for any property F. Put a little paradoxically, "If two things are the same, anything you do to them both will give the same answer". For example, if $a = b$, then $\{a, c\} = \{b, c\}$ for any c.
3. The aim of the axioms is to incorporate, in a few short statements, everything that different vector spaces have in common. Everything that is true of all vector spaces is a theorem that follows from the axioms. The significance of axioms will be discussed again in Chapter 15.

Subspaces

It sometimes happens that one vector space is inside, or part of, another. We say that the smaller is a *subspace* of the larger. For example, the vector space of polynomials of degree at most 3, is a subspace of the vector space of all polynomials.

It is often important to know whether a subset (or part, W, of a given vector space V) is a subspace of V, that is, whether the subset, considered by itself, forms a vector space. The main question is whether W forms a system *in* which addition and multiplication can be carried out: adding two members of the subset W must give a vector which is again in W (and not just anywhere in V), and similarly for scalar multiplication. Thus, for a (non-empty) subset W of a vector space V to be a *subspace* of V, we must have:

1. for all $\underset{\sim}{u}, \underset{\sim}{v} \in W, \quad \underset{\sim}{u} + \underset{\sim}{v} \in W$
   (we say "W is closed under addition"); and
2. for all $\gamma \in \mathbf{R}$ and $\underset{\sim}{u} \in W, \quad \gamma\underset{\sim}{u} \in W$
   (we say "W is closed under scalar multiplication").

Consider, as an example, the set W of all the vectors in the xy-plane. This is the plane $z = 0$, that is, it consists of all the vectors whose last co-ordinate is zero. So every vector in W is of the form (x, y, 0) for some $x, y \in \mathbf{R}$:

$$W = \{(x, y, 0): x, y \in \mathbf{R}\}$$

Firstly, we know that $W \subset \mathbf{R}^3$ (and of course it is not empty). For any two vectors,

$$\underset{\sim}{u} = (a_1, a_2, 0)$$

and,

$$\underset{\sim}{v} = (b_1, b_2, 0)$$

in this set we have:

$$\begin{aligned}\underset{\sim}{u} + \underset{\sim}{v} &= (a_1 + b_1, a_2 + b_2, 0)\\ &\in W\end{aligned}$$

(since its last coordinate is zero)

and,

$$\begin{aligned}\gamma\underset{\sim}{u} &= \gamma(a_1, a_2, 0)\\ &= (\gamma a_1, \gamma a_2, 0)\\ &\in W\end{aligned}$$

(since its last co-ordinate is zero).

Hence, W is a subspace of $\mathbf{R}^3$. ▌

The closure of the xy-plane under addition and scalar multiplication is easily interpreted geometrically, as shown in Figures 9.1 and 9.2.

If $\underset{\sim}{u}$ and $\underset{\sim}{v}$ are vectors in the xy-plane (i.e. their endpoints, and therefore the whole arrows, lie in the xy-plane), it is clear that $\underset{\sim}{u} + \underset{\sim}{v}$ also lies in the xy-plane (Figure 9.1).

If $\underset{\sim}{u}$ lies in the xy-plane, then any multiple of it, $\gamma\underset{\sim}{u}$, also lies in the xy-plane (Figure 9.2). This is useful as a "standard example" of a subspace. When the word "subspace"

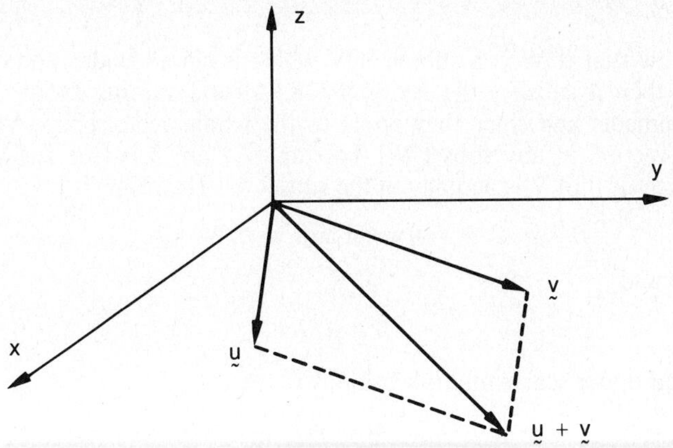

Figure 9.1

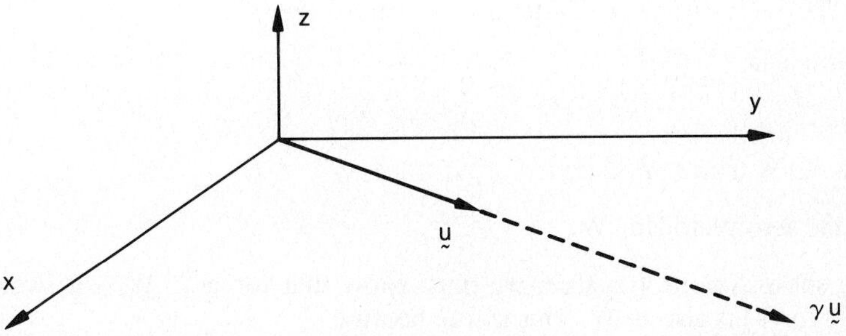

Figure 9.2

comes up, do not think "satisfies some rules that are in my notes somewhere", but think instead "is like the xy-plane in $\mathbf{R}^3$".

Notice that to prove that a subset W *is* a vector space, we had to prove two "all" statements:

$$\text{For } all\ \underset{\sim}{u},\ \underset{\sim}{v} \in W,\quad \underset{\sim}{u} + \underset{\sim}{v} \in W$$

and,

$$\text{For } all\ \underset{\sim}{u} \in W \text{ and } all \text{ scalars } \gamma \in \mathbf{R},\quad \gamma\underset{\sim}{u} \in W$$

This is why, in the above example, we took,

$$\underset{\sim}{u} = (a_1, a_2, 0),\ \underset{\sim}{v} = (b_1, b_2, 0)$$

and a general γ; it would not be sufficient to take particular examples for $\underset{\sim}{u}$, $\underset{\sim}{v}$ and γ, say $\underset{\sim}{u} = (1, 3, 0)$, $\underset{\sim}{v} = (4, 3, 0)$ and $\gamma = 2$.

We have said that a subspace of a vector space must be closed under addition and scalar multiplication. Now in fact the converse is also true:

Theorem

If a (non-empty) subset W of a vector space is closed under addition and scalar multiplication, then it is a subspace of V.

Proof

We have to show that if W is a subset of V which is closed under addition and scalar multiplication, then it satisfies the Axioms 1–8. All the axioms except Axioms 3 and 4 are "all" statements and since they apply to the whole vector space V, they are true also of all the vectors in any subset W. To show Axiom 3 is true for W, we have to show that the vector $\underset{\sim}{0}$ of V is actually in the subset W. However,

$$\underset{\sim}{0} = 0\underset{\sim}{w} \text{ for any } \underset{\sim}{w} \in W$$

(proved above) and,

$$0\underset{\sim}{w} \in W$$

since W is closed under scalar multiplication, so,

$$\underset{\sim}{0} \in W.$$

The definition of $\underset{\sim}{0}$ was,

$$\underset{\sim}{0} + \underset{\sim}{v} = \underset{\sim}{v} \text{ for all } \underset{\sim}{v} \in V$$

so in particular,

$$\underset{\sim}{0} + \underset{\sim}{w} = \underset{\sim}{w}$$

for all $\underset{\sim}{w} \in W$ (since $W \subset V$).

So $\underset{\sim}{0}$ is the zero vector for W.

To show Axiom 4 is true, we must show that for $\underset{\sim}{u} \in W$, $-\underset{\sim}{u}$ (which exists by Axiom 4 for V) is also in W. This is true because,

$$-\underset{\sim}{u} = (-1)\underset{\sim}{u}$$

which is in W since W is closed under scalar multiplication. (In fact, we have not yet proved that,

$$-\underset{\sim}{u} = (-1)\underset{\sim}{u}$$

This is left to be attempted in Exercise 3(a)).

Thus W satisfies Axioms 1–8, so W is a subset of V which is itself a vector space.

That is, W is a subspace of V. ▌

Note

Where have we used the assumption that W is closed under addition? W is a vector space if it has an addition and scalar multiplication satisfying Axioms 1 to 8. If W were not closed under addition, it would not "have an addition" at all; W would not be a system in which we could add elements, so we would not even start to ask if the axioms were true.

From the result just proved, we can derive the following rule:

To prove that a given subset W of a given vector space V is a subspace of V, we have to prove that W is closed under addition and scalar multiplication. In other words, to show a subset of V *is* a subspace of V, we must prove two "all" statements.

The situation is different, however, when we wish to prove that a given subset of V is *not* a subspace of V. We only have to show that one of the "all" statements is false, and to show that an "all" statement is false, we need exhibit only one counterexample. (This was discussed in Chapter 6.) Thus, to show a subset W is *not* a subspace, exhibit *one* counterexample to *one* of the rules.

Example 1

Show that,

$$W = \{(x, y) \in \mathbf{R}^2 \mid x + y = 1\}$$

is not a subspace of $\mathbf{R}^2$.

Proof

$(\frac{1}{2}, \frac{1}{2})$ and $(1, 0)$ are in W but their sum is,

$$(\tfrac{1}{2}, \tfrac{1}{2}) + (1, 0) = (1\tfrac{1}{2}, \tfrac{1}{2})$$

which is not in W.

So W is not closed under addition.

So W is not a subspace of $\mathbf{R}^2$. ▌

The counterexample is easily understood with the help of a diagram. See Figure 9.3.

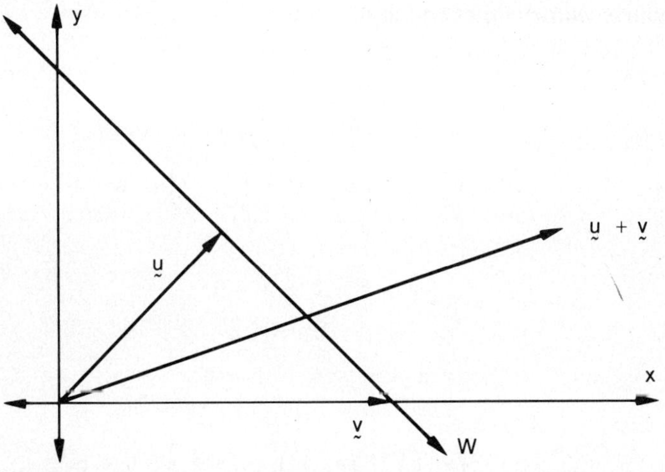

Figure 9.3

Note:
The points $\underset{\sim}{u} = (\frac{1}{2}, \frac{1}{2})$ and $\underset{\sim}{v} = (1, 0)$ are in W, but their sum $\underset{\sim}{u} + \underset{\sim}{v} = (1\frac{1}{2}, \frac{1}{2})$ is not in W.

Given the question, "Is a given subset W of V a subspace or not? Prove your answer", it will be necessary to either choose a few vectors and see if a counterexample appears, or to attempt to prove closure under addition and scalar multiplication. However, a failure to carry through such a proof is not a proof that W is *not* a subspace. The following old Middle Eastern story illustrates this point. An old man lived with his wife by the side of the River Tigris. One day his wife found him lost in thought. When she asked why, he answered that he had just concluded that his strength had not changed since the day they had moved there, 40 years earlier. His wife, knowing this was not true, asked how he came to this conclusion. "I couldn't move this boulder then," he said, "and I still can't move it now. So my strength hasn't changed."

It must be emphasised that a particular counterexample to one of the two rules must be found to prove that W is not a subspace.

It turns out that most subsets of a vector space are not subspaces.

Some further useful techniques of proof are illustrated by the following example.

Example 2

Show that,

$$W = \left\{ \begin{pmatrix} a & b \\ c & d \end{pmatrix} \in M_{22}: a + b + c + d = 0 \right\}$$

is a subspace of M_{22}.

Proof

A matrix is in W if and only if it is 2 × 2 and the sum of its entries is 0. (Recall from Chapter 7 that $y \in \{x: Fx\}$ just means Fy.) So we have to show that if we add two matrices whose entries add to 0, or multiply such a matrix by a real number, the result is again a matrix whose entries add to 0.

Method 1

Let,

$$A = \begin{pmatrix} a & b \\ c & d \end{pmatrix} \quad \text{and} \quad A' = \begin{pmatrix} a' & b' \\ c' & d' \end{pmatrix} \in W$$

so that,

$$a + b + c + d = 0$$

and,

$$a' + b' + c' + d' = 0$$

then,

$$A + A' = \begin{pmatrix} a & b \\ c & d \end{pmatrix} + \begin{pmatrix} a' & b' \\ c' & d' \end{pmatrix} = \begin{pmatrix} a + a' & b + b' \\ c + c' & d + d' \end{pmatrix}.$$

Now,

$$\begin{aligned} (a + a') + (b + b') + (c + c') + (d + d') &= (a + b + c + d) + (a' + b' + c' + d') \\ &= 0 + 0 \\ &= 0. \end{aligned}$$

Therefore,

$$A + A' \in W$$

so we have shown that W is closed under addition.

Let,

$$A = \begin{pmatrix} a & b \\ c & d \end{pmatrix} \in W$$

so that,

$$a + b + c + d = 0$$

and let α be a scalar.

$$\alpha A = \alpha \begin{pmatrix} a & b \\ c & d \end{pmatrix} = \begin{pmatrix} \alpha a & \alpha b \\ \alpha c & \alpha d \end{pmatrix}$$

and,

$$\begin{aligned} \alpha a + \alpha b + \alpha c + \alpha d &= \alpha(a + b + c + d) \\ &= \alpha 0 \\ &= 0 \end{aligned}$$

So,

$$\alpha A \in W$$

Therefore W is a subspace of M_{22}.

Method 2

If,

$$A = \begin{pmatrix} a & b \\ c & d \end{pmatrix} \in W$$

then,

$$a + b + c + d = 0$$

so,

$$d = -a - b - c$$

So,

$$A = \begin{pmatrix} a & b \\ c & -a - b - c \end{pmatrix}$$

This gives a "general form" for W, that is, W consists of all the matrices of this form.

Let,

$$A, A' \in W, \quad \text{so } A = \begin{pmatrix} a & b \\ c & -a - b - c \end{pmatrix}, \; A' = \begin{pmatrix} a' & b' \\ c' & -a' - b' - c' \end{pmatrix}$$

So,

$$\begin{aligned} A + A' &= \begin{pmatrix} a + a' & b + b' \\ c + c' & -a - b - c - a' - b' - c' \end{pmatrix} \\ &= \begin{pmatrix} a + a' & b + b' \\ c + c' & -(a + a') - (b + b') - (c + c') \end{pmatrix} \end{aligned}$$

which is of the same form, and so is in W.

So W is closed under addition.

Let,

$$A \in W, \text{ so } A = \begin{pmatrix} a & b \\ c & -a - b - c \end{pmatrix}$$

and let α be a scalar.

Then,

$$\alpha A = \begin{pmatrix} \alpha a & \alpha b \\ \alpha c & \alpha(-a - b - c) \end{pmatrix}$$

$$= \begin{pmatrix} \alpha a & \alpha b \\ \alpha c & -\alpha a - \alpha b - \alpha c \end{pmatrix}$$

which is of the same form, and so is in W.

So W is closed under scalar multiplication and is therefore a subspace of M_{22}. ▌

(Note that in both methods we have applied the techniques of Chapter 2 for proving "all" statements.)

The second method, which uses a general form, has the advantage of making clear exactly what the elements of the subspace look like. However, a general form is sometimes impossible to find, or is too complicated (see Exercise 8 for an example) and therefore the first method is more general.

It is possible to develop the various notions of linear algebra—linear combinations, linear dependence and independence, span, basis and dimension, coordinates, linear transformations—in the general setting of vector space theory. This can be done because these notions are all defined in terms of addition, scalar multiplication and general logical and set-theoretical concepts. Enough has been said here and in earlier chapters to indicate how to deal with definitions and proofs involving these notions.

Exercises

(Gradings of exercises: * easy, ** moderate, *** difficult.)

Vector spaces

*1. Show that the set of positive integers is not a vector space.

**2. Show that $\mathbf{R}^2$ is a vector space. You must check Axioms 1–8. Addition is defined by, $(x_1, y_1) + (x_2, y_2) = (x_1 + x_2, y_1 + y_2)$ and scalar multiplication by, $\alpha(x, y) = (\alpha x, \alpha y)$. You may assume the laws of real numbers such as the associative law:

$$(x + y) + z = x + (y + z) \text{ for all } x, y, z \in \mathbf{R}$$

***3. Prove from the axioms that in a vector space:

(a) $(-1)\underset{\sim}{u} = -\underset{\sim}{u}$ for any vector $\underset{\sim}{u}$

(You may use the result $0\underset{\sim}{u} = \underset{\sim}{0}$, which was proved in the text.)

(b) For all vectors $\underset{\sim}{u}, \underset{\sim}{v}, \underset{\sim}{w}$, if $\underset{\sim}{u} + \underset{\sim}{w} = \underset{\sim}{v} + \underset{\sim}{w}$ then $\underset{\sim}{u} = \underset{\sim}{v}$.

(Perhaps these facts are at least as obvious as the axioms themselves. The point of the proofs here is to derive the results, at each step, using only an axiom and not relying on any concepts of what is obvious.)

Subspaces

***4.** Prove that, $S = \{(x, y) \in \mathbf{R}^2 : x - y = 1\}$ is not a vector space.

***5.** Prove that,

$$W = \left\{\begin{pmatrix} a & b \\ c & d \end{pmatrix} \in M_{22} : c - d = 0\right\}$$

is a subspace of M_{22}.

***6.** Prove that the following subset of $\mathbf{R}^3$ is *not* a subspace,

$$\{(a, b, c) : a + b + c = 1\}$$

***7.** The set $W = \{(x, 0, y) : x, y \in \mathbf{R}\}$ is a subset of $\mathbf{R}^3$.
(a) Prove that W is a subspace of $\mathbf{R}^3$.
(b) Is W a vector space? Explain your answer briefly.
(c) Interpret W geometrically.

****8.** Let $A \in M_{22}$. Show that, $\{X \in M_{22} : AX = XA\}$ is a subspace of M_{22}.

****9.** (a) Prove that if W is a subspace of a vector space V then W contains the zero vector.
(b) Prove that the converse of this statement is not true.

*****10.** Let C be the vector space consisting of all continuous functions from **R** to **R**. (Addition is defined by, $(f + g)(x) = f(x) + g(x)$ and scalar multiplication by, $(\alpha f)(x) = \alpha f(x)$. Show that:
(a) $W = \{f \in C : \int_0^1 f(x)\,dx = 0\}$ is a subspace of C.
(b) $H = \{f \in C : \int_0^1 f(x)\,dx = 1\}$ is not a subspace of C.

10

Limits and convergence

One of the main applications of the theory of multiple quantifiers, which was outlined in Chapter 5, is in analysing the notions of limit and convergence. These concepts arise in two main areas, in calculus and in the theory of computation. In calculus, they are needed to make precise the notions of limit at infinity and of continuity, and, in fact, are needed to make precise all the fundamental ideas of calculus. In the theory of computation, they are needed in studying the central topic of the accuracy of approximate calculations. The concepts of limit and convergence are often found to be difficult. The difficulties are mainly logical.

Limit of a function "at infinity"

To say that the function,

$$f(x) = \frac{x + 1}{x}$$

approaches 1 as x approaches infinity, or in symbols,

$$f(x) \to 1 \quad \text{as} \quad x \to \infty$$

or,

$$\lim_{x \to \infty} f(x) = 1,$$

means that as x becomes larger and larger, $(x + 1)/x$ gets closer and closer to 1. Some values have been tabulated in Table 10.1.

Table 10.1

x	1	10	500	1 million
$\frac{x+1}{x}$	2	1.1	1.002	1.000001

By taking x large enough, (x + 1)/x can be made "arbitrarily close" or "as close as we please" to 1. In other words, given any desired accuracy, say 0.001, we can find how large x has to be so that (x + 1)/x is within 0.001 of 1. Note also that for all larger values of x, (x + 1)/x stays within 0.001 of 1, as shown in Figure 10.1. In general, for any positive number ϵ that we choose, however small, we can take x large enough so that, from that point on, (x + 1)/x is wihin ϵ of 1. (The small Greek letter ϵ, epsilon, is traditionally used for an arbitrary small positive number.) The phrases "that we choose", "large enough", "as close as we please", are used to facilitate understanding of a notion that is purely logical. The final definition of,

$$\lim_{x \to \infty} \frac{x + 1}{x} = 1$$

is,

$$\text{For all } \epsilon > 0 \text{ there exists X such that for all } x > X, \left| \frac{x + 1}{x} - 1 \right| < \epsilon$$

In general,

$$\lim_{x \to \infty} f(x) = L$$

means,

$$\text{For all } \epsilon > 0 \text{ there exists X such that for all } x > X, |f(x) - L| < \epsilon$$

Notes

1. Remember that $|x - y| < \epsilon$ can be thought of as saying that the distance between x and y is less than ϵ. In the example above, the mod signs could be left out, since, in fact, (for x positive) (x + 1)/x is always *greater* than 1. But in other cases f(x) could be sometimes greater than the limit and sometimes less.
2. The symbol ∞, for infinity, is not meant to name any number, as if (x + 1)/x actually reaches

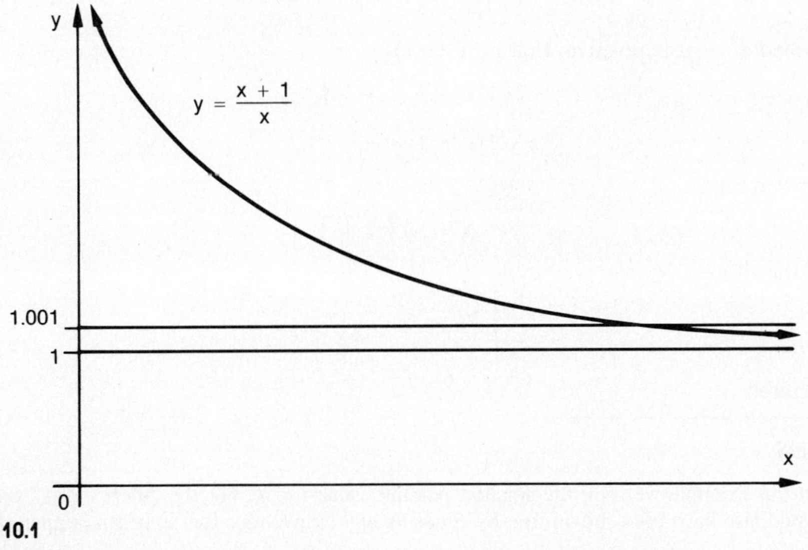

Figure 10.1

1 when x "equals infinity". It is just a shorthand way of expressing what happens for larger and larger (finite) numbers.
3. The point of the last quantifier in the definition, "for all $x > X \ldots$", is to rule out the case of a function which might move close to a number for some x but then move away again. See Exercise 2 at the end of this chapter.

Example 1

Prove that,

$$\lim_{x\to\infty} \frac{2x^2 - 3}{x^2 - 1} = 2$$

Proof

Let $\epsilon > 0$.
We wish to find X such that for all $x > X$,

$$\left| \frac{2x^2 - 3}{x^2 - 1} - 2 \right| < \epsilon$$

Now,

$$\left| \frac{2x^2 - 3}{x^2 - 1} - 2 \right| = \left| \frac{2x^2 - 3 - 2(x^2 - 1)}{x^2 - 1} \right|$$

$$= \left| \frac{-1}{x^2 - 1} \right|$$

which is $\dfrac{1}{x^2 - 1}$ for large x.

We solve,

$$\frac{1}{x^2 - 1} < \epsilon$$

$$1/\epsilon < x^2 - 1$$

(provided $x^2 - 1$ is positive, that is, $x > 1$)

$$1/\epsilon + 1 < x^2$$

$$\sqrt{1/\epsilon + 1} < x$$

So we can take,

$$X = \sqrt{1/\epsilon + 1}$$

Then for all $x > X$,

$$\left| \frac{2x^2 - 3}{x^2 - 1} - 2 \right| < \epsilon$$

as required.

Notes

1. In this example we took the smallest possible value for X, but the "there exists" statement could also have been shown true by choosing any larger value for X. It can happen that it is

easy to find a possible value for X, but hard to find the least. See Exercise 2(c) at the end of this chapter.

2. One step in the proof only applied provided that x was greater than 1. Of course, we were only thinking of large numbers x, but to make sure there are no gaps in the reasoning, we should check that the number we choose for X is greater than 1. In this case it is, but if it had not been, we could have chosen X to be the *maximum* of $\sqrt{1/\epsilon + 1}$ and 1.
3. In accordance with the techniques of Chapter 5, to show a, "For all ϵ there exists X . . ." statement, we take an arbitrary ϵ and then find X (satisfying the required condition) *in terms of* ϵ. A different ϵ would require, usually, a different choice of X. In this example, the smaller ϵ is, the larger X will have to be.

Limit of a sequence

Essentially the same ideas, involving the logical analysis of the requirements for achieving a given accuracy, appear in other contexts, both continuous and discrete. For example, the sequence,

$$\frac{1}{2}, \frac{2}{3}, \frac{3}{4}, \ldots, \frac{n}{n+1}, \ldots$$

approaches 1, or has limit 1. For a sequence,

$$a_1, a_2, a_3, \ldots, a_n, \ldots$$

the assertion,

$$\lim_{n \to \infty} a_n = L$$

may be analysed as:

For all $\epsilon > 0$, there exists a whole number N such that for all $n > N$, $|a_n - L| < \epsilon$.

Example 2

Prove that,

$$\lim_{n \to \infty} \frac{(-1)^n}{n^2} = 0$$

Proof

We have to show that for all $\epsilon > 0$ there exists a whole number N such that for all $n > N$,

$$\left| \frac{(-1)^n}{n^2} - 0 \right| < \epsilon$$

That is, we want,

$$\frac{1}{n^2} < \epsilon$$

to hold. But this will hold if and only if,

$$n^2 > \frac{1}{\epsilon}$$

and this will hold if and only if,

$$n > \frac{1}{\sqrt{\epsilon}}$$

We can therefore choose N to be the first integer greater than or equal to $\frac{1}{\sqrt{\epsilon}}$ ∎

To connect this piece of logic with intuition, let us take as an example,

$$\epsilon = \frac{1}{1000}$$

Then,

$$\frac{1}{\sqrt{\epsilon}} = 31.6$$

We would therefore choose N to be 32.

Choosing N = 32 we will ensure that for all n > N, $\left| \frac{(\ 1)^n}{n^2} - 0 \right| < \epsilon.$

See Figure 10.2.

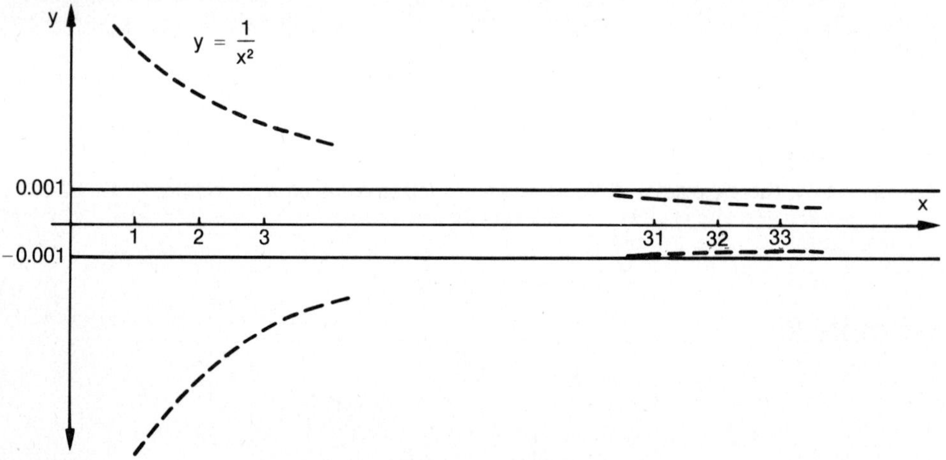

Figure 10.2

Continuity

Calculus is the science of the continuous, so a serious treatment of calculus requires a precise definition of continuity. Intuitively, a function is continuous if its graph is joined up, as illustrated in Figure 10.3. A first step in analysing this notion is to observe that for continuous functions, small changes in x result in only small changes in the function value. For non-continuous functions, this is not true: sometimes a small change in x results in a large change, or jump, in the function.

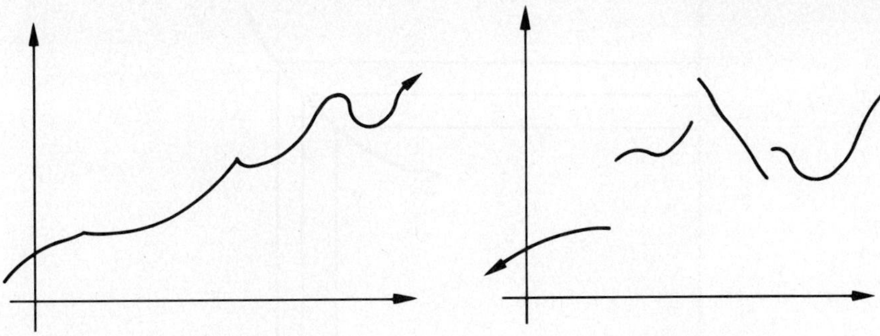

Figure 10.3 (a) *Continuous* (b) *Not continuous*

Next, it is necessary to say what it means for a function to be continuous at a point a, that is, to have no jump at a. This is so if, for values of x near a, f(x) is near f(a), as illustrated in Figure 10.4. Now, if we ask "How near?", we realise that the point is to be able to get "as near as we please" to f(a) by making x near enough to a.If there were a jump in the graph of the function, f(x) would not get close to f(a) as x got close to a. So for any $\epsilon > 0$ (however small) the values of f must be all within ϵ of f(a) for some distance around a. Using a "there exists" statement to translate the "some", the final expression is:

"f(x) is continuous at a"

means:

For all $\epsilon > 0$ there exists $\delta > 0$ such that, whenever $|x - a| < \delta$, $|f(x) - f(a)| < \epsilon$

See Figure 10.5.

A function is then *continuous* if it is continuous at all points a. So for the definition of "f is continuous" we just add "for all real numbers a" to the beginning of the last definition. So the definition of "f is continuous" is:

For all the real numbers a, for all $\epsilon > 0$, there exists $\delta > 0$
Such that whenever $|x - a| < \delta$, $|f(x) - f(a)| < \epsilon$

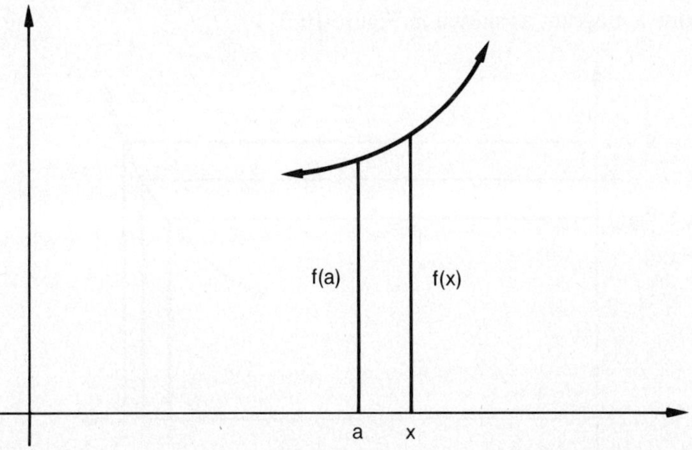

Figure 10.4

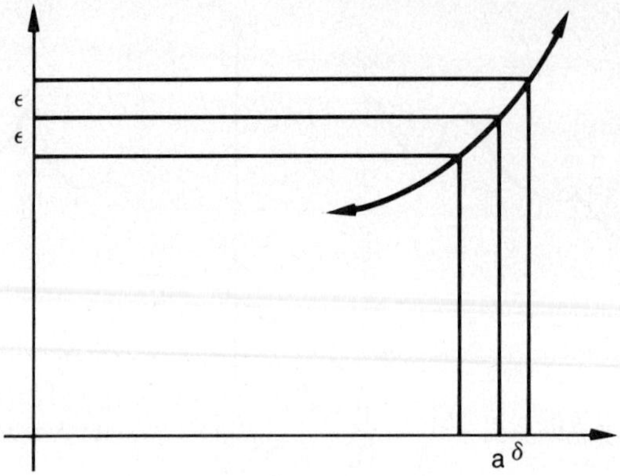

Figure 10.5

Note:
The δ will be smaller for smaller ϵ, and will also be smaller the steeper the graph of f.

Example 3

Show that the function, $f(x) = x^2$ is continuous at $x = 3$.

(Why *prove* such an intuitively obvious result? Firstly, practice on an intuitively clear example allows us to check that our formal definition *does* accord with intuition. Secondly, practice with the formal definition in a simple case is necessary before going on to more complicated cases, where intuition cannot be trusted. As will be described in Chapter 15, the logical definitions of calculus concepts were developed after mistakes were made through relying on geometrical intuition).

Proof

We have to show that for all $\epsilon > 0$, there exists a positive number δ such that,

$$|f(x) - 9| < \epsilon$$

whenever,

$$0 < |x - 3| < \delta$$

Let $\epsilon > 0$.

Then we draw a diagram as shown in Figure 10.6.

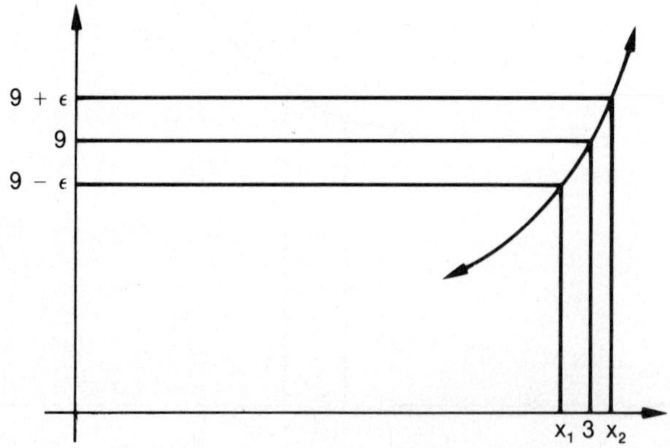

Figure 10.6

Now,

$$\delta_2 = x_2 - 3 = \sqrt{9 + \epsilon} - 3$$

and,

$$\delta_1 = 3 - x_1 = 3 - \sqrt{9 - \epsilon}$$

We take δ as the smaller of δ_1 and δ_2 (in fact the smaller is $\delta_2 = \sqrt{9 + \epsilon} - 3$ but here we do not need to know this).

Then any x satisfying,

$$0 < |x - 3| < \delta$$

will ensure that x^2 will differ from 9 by an amount less than ϵ.
For if $|x - 3| < \delta$ then,

$$3 - (3 - \sqrt{9 - \epsilon}) < x < 3 + (\sqrt{9 + \epsilon} - 3)$$
$$\sqrt{9 - \epsilon} < x < \sqrt{9 + \epsilon}$$
$$9 - \epsilon < x^2 < 9 + \epsilon$$
$$|x^2 - 9| < \epsilon \qquad \blacksquare$$

For example if $\epsilon = 0.1$ then a suitable choice is:

$$\delta = \sqrt{9.1} - 3 = 0.0166$$

This method gives the maximum possible value of δ that will work. Any other smaller positive value will also work.

Another approach to proving that $f(x) = x^2$ is continuous at $x = 3$ is as follows:

The derivative of $y = x^2$ at $x = 3$ is 6; this means that (near $x = 3$) a small change δ in x should cause a change of about 6δ in x^2, so that a change of ϵ in x^2 should be caused by a change of about $\epsilon/6$ in x.

So let $\epsilon > 0$ and take something smaller than $\epsilon/6$ for δ; say $\epsilon/7$.

If,

$$|x - 3| < \frac{\epsilon}{7}$$

$$3 - \frac{\epsilon}{7} < x < 3 + \frac{\epsilon}{7}$$

$$9 - \frac{6\epsilon}{7} + \left(\frac{\epsilon}{7}\right)^2 < x^2 < 9 + \frac{6\epsilon}{7} + \left(\frac{\epsilon}{7}\right)^2$$

Now,

$$9 - \epsilon < 9 - \frac{6\epsilon}{7} < 9 - \frac{6\epsilon}{7} + \left(\frac{\epsilon}{7}\right)^2$$

so,

$$9 - \epsilon < x^2$$

also, provided $\frac{\epsilon}{7} < 1$,

$$\left(\frac{\epsilon}{7}\right)^2 < \frac{\epsilon}{7}$$

so

$$9 + \frac{6\epsilon}{7} + \left(\frac{\epsilon}{7}\right)^2 < 9 + \frac{6\epsilon}{7} + \frac{\epsilon}{7}$$
$$= 9 + \epsilon$$

This has shown that, if $\epsilon/7 < 1$, then $|x - 3| < \epsilon/7$ implies $|x^2 - 9| < \epsilon$.

So if we now take δ = minimum of $\epsilon/7$ and 1, then if $|x - 3| < \delta$ then $|x^2 - 9| < \epsilon$.

Therefore, $f(x) = x^2$ is continuous at $x = 3$. ▮

Calculating

To calculate such quantities as sin x, e^x and log x, a computer or calculator must represent these functions as combinations of the elementary operations (addition, subtraction, multiplication and division). The functions may be represented exactly by *infinite* combinations of these operations; for example,

$$e^x = 1 + x + \frac{x^2}{2!} + \frac{x^3}{3!} + \ldots + \frac{x^n}{n!} + \ldots$$

A calculator or computer cannot, of course, calculate infinitely many terms. However, the series that are used, such as the one for e^x, have the property of being *convergent*—for any x, adding up some finite number of terms approximates the answer to any desired accuracy. The analysis of this notion involves the same ideas as those given above under the topics of limits and continuity. For example, to say that the above series converges to e^x for all x means:

For all x, for all $\epsilon > 0$, there exists N such that for all $n > N$,

$$\left| e^x - \left(1 + x + \frac{x^2}{2!} + \ldots + \frac{x^n}{n!}\right)\right| < \epsilon$$

N depends on x and ϵ—the larger x is, and the smaller the desired margin of error ϵ, the larger N will have to be.

The divergence of the harmonic series

If we keep adding,

$$\frac{1}{2} + \frac{1}{4} + \frac{1}{8} + \frac{1}{16} + \ldots + \frac{1}{2^n} + \ldots$$

we get closer and closer to 1. Using the concepts above, we could make this precise by saying: For all $\epsilon > 0$ there exists N such that for all $n > N$,

$$1 > \frac{1}{2} + \frac{1}{4} + \frac{1}{8} + \ldots + \frac{1}{2^n} > 1 - \epsilon$$

(Note that there are no dots after the $1/2^n$, because here we are speaking of a finite sum.) This is sometimes expressed by saying, "the sum to infinity of the series is 1" or, "the series converges to 1". This fact is obvious from the diagram of a 1×1 square in Figure 10.7. A diagram is not considered a proof (which is a little unfair, since no proof could be more convincing than this diagram). A proof would normally begin with the formula for the sum of a geometric progression.

What about the series,

$$\frac{1}{2} + \frac{1}{3} + \frac{1}{4} + \frac{1}{5} + \frac{1}{6} + \frac{1}{7} + \ldots + \frac{1}{n} + \ldots ?$$

This is called the *harmonic* series. This does not look very different in principle to the series above and one might expect it to add up to some number (obviously bigger than 1, since, $1/2 + 1/3 + 1/4$ is already bigger than 1). The fact is, however, that the harmonic series *diverges*, or grows indefinitely. This means that by adding enough terms in the series, we can get past any number, however big. In purely logical terms: For any real number M there exists n such that,

$$\frac{1}{2} + \frac{1}{3} + \frac{1}{4} + \ldots + \frac{1}{n} > M$$

The proof is set out below. It is another of the classic proofs of mathematics and has a quite clever trick in it, but is not hard to understand.

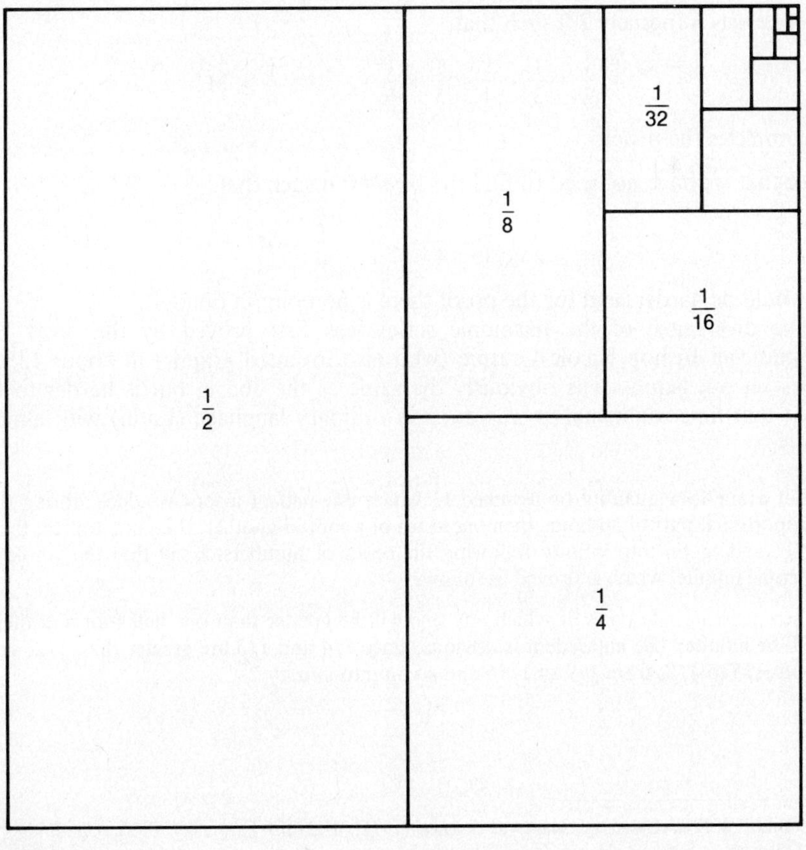

Figure 10.7

Proof

Let M be a real number.

Now for any whole number m,

$$\frac{1}{2}+\frac{1}{3}+\frac{1}{4}+\frac{1}{5}+\frac{1}{6}+\frac{1}{7}+\frac{1}{8}+\frac{1}{9}+\frac{1}{10}+\ldots+\frac{1}{16}+\frac{1}{17}+\ldots+\frac{1}{32}+\ldots+\frac{1}{2^m}$$

$$>\frac{1}{2}+\frac{1}{4}+\frac{1}{4}+\frac{1}{8}+\frac{1}{8}+\frac{1}{8}+\frac{1}{8}+\frac{1}{16}+\frac{1}{16}+\ldots+\frac{1}{16}+\frac{1}{32}+\ldots+\frac{1}{32}+\ldots+\frac{1}{2^m}$$

(because each term in the top line is greater than or equal to the term written below it).

$$=\frac{1}{2}+\frac{1}{2}+\frac{1}{2}+\frac{1}{2}+\frac{1}{2}+\ldots+\frac{1}{2}$$

(with m $\frac{1}{2}$'s).

It is possible to take m large enough so that,

$$m \times \frac{1}{2} > M$$

(m can be any integer greater than 2M).

So there exists n (namely 2^m) such that,

$$\frac{1}{2}+\frac{1}{3}+\frac{1}{4}+\frac{1}{5}+\frac{1}{6}+\ldots+\frac{1}{n} > M$$

This completes the proof.

(Notice that we have not tried to find the *smallest* n such that,

$$\frac{1}{2}+\frac{1}{3}+\frac{1}{4}+\ldots+\frac{1}{n} > M$$

That would be harder, and for the proof there is no point in doing it.)

The divergence of the harmonic series was first proved by the great French mathematician Bishop Nicole Oresme (who also invented graphs) in about 1350. His proof is set out below. It is obviously the same as the above, but is harder to follow since at that time mathematics was done in ordinary language (Latin) with almost no symbols.

> "Let a one-foot quantity be assumed to which one-half of a foot is added during the first proportional part of an hour, then one-third of a foot in another, then one-fourth, then one-fifth, and so on into infinity following the series of numbers. I say that the whole would become infinite, which is proved as follows:
>
> There exist infinite parts of which any one will be greater than one-half foot and the whole will be infinite. The antecedent is obvious, since 1/4 and 1/3 are greater than 1/2; similarly from 1/5 to 1/8, from 1/9 to 1/16 and so on into infinity."[1]

---

1. E. Grant, *A Sourcebook of Medieval Science* (1974), Harvard University Press, Cambridge, Mass., p. 135.

Exercises

(Gradings of exercises: * easy, ** moderate, *** difficult.)

Limit of a function at infinity

**1. (a) What is:

$$\lim_{x \to \infty} \frac{x^2 + 1}{x^2}$$

(If you do not know any technique for this, use a calculator to evaluate $\frac{x^2 + 1}{x^2}$ for some large values of x, and guess the answer.)

(b) Find an X such that for all $x > X$, $\frac{x^2 + 1}{x^2}$ is within 0.01 of its limit.

(c) For any $\epsilon > 0$, find an X such that for all $x > X$, $\frac{x^2 + 1}{x^2}$ is within ϵ of its limit.

**2. Answer the same questions as in Question 1 for $\frac{\sin x}{x}$.

Find, approximately, the first $X > 0$ such that $\frac{\sin X}{X}$ is within 0.01 of $\lim_{x \to \infty} \frac{\sin x}{x}$.

Is it true that, for *this* X, $\frac{\sin x}{x}$ is within 0.01 of its limit for all $x > X$?

**3. One way of giving hospital patients oxygen is to cover their beds with small tents into which the gas is pumped. Wayne and Chamney, in a 1969 paper discussing oxygen tent performance, used the following equation for the concentration of oxygen, x(t), where t is the time in minutes after starting the oxygen inflow:

$$x(t) = \frac{1}{f + E}(f + 0.21E - 0.79f\exp(-(f + E)t/V))$$

where E is the tent gas exchange rate (that is, rate of leakage in and out), f the oxygen inflow rate and V the tent gas volume; these are all constants for a given tent.

(a) What is the "steady state" or "equilibrium state" (that is, what happens to the concentration as $t \to \infty$)?

(b) On the first trial, f was 18.0 litres/min, E was 17.0 litres/min and V was 145 litres. What was the equilibrium concentration? How long would it take after starting the inflow for the concentration to reach within 0.01 of its equilibrium state?

D. J. Wayne and A. R. Chamney, 'Oxygen Tent Performance', *Physics in Medicine and Biology*, (1969), **14**, 9–17.

**4. Prove that:

$$\lim_{x \to \infty} \frac{x - 1}{2x + 1} = \frac{1}{2}.$$

Limit of a sequence

*5. What is meant by, "A sequence $\{a_n\}$ has a limit b"?

*6. (a) What would you do to prove that:

$$\lim_{n \to \infty} \frac{1}{n} = 0?$$

(b) Prove that:

$$\lim_{n \to \infty} \frac{1}{n} = 0.$$

(c) Given $\epsilon = .0001$, produce an integer N such that for all $n > N$,

$$\left| \frac{1}{n} - 0 \right| < \epsilon$$

*7. Prove that:

$$\lim_{n \to \infty} \frac{n + 1}{2n} = \frac{1}{2}$$

**8. The nth term a_n of a sequence is given by:

$$a_n = \frac{1 - n}{3n + 5}$$

(a) Write down the 1,000th, 10,000th and 1,000,000th terms of the sequence.
(b) Make a reasonable guess for $\lim_{n \to \infty} a_n$
(c) Prove your answer.

**9. (a) Prove that:

$$\lim_{n \to \infty} (1)^n = 1$$

(b) Prove that $\lim_{n \to \infty} (-1)^n$ does not exist.

**10. Prove that:

$$\lim_{n \to \infty} \frac{3n^2 - 1}{2n^2 + n + 1} = \frac{3}{2}$$

**11. (a) Prove that:

$$\lim_{n \to \infty} \left(\frac{1}{2^n}\right) = 0$$

(b) Prove that $\lim_{n \to \infty} 2^n$ does not exist.

**12. Prove that $\lim_{n \to \infty} \sin n$ does not exist.

Continuity

***13.** Let,

$$f(x) = x^2 \text{ and } \epsilon = \frac{1}{2}$$

Find a $\delta > 0$ such that:

$$|f(x) - 4| < \epsilon$$

whenever,

$$|x - 2| < \delta$$

***14.** A function is given by, $f(x) = 4x + 3$. Show that f is continuous at $x = 3$.

****15.** Prove that the function f given by $f(x) = \frac{1}{x}$ is continuous at $x = 2$.

****16.** A function f is given by $f(x) = x^2$ for all x. Prove that:
(a) f is continuous at $x = 4$.
(b) f is continuous at all x.

****17.** A function f is given by:

$$f(x) = \begin{cases} \frac{x}{x} & \text{for } x \neq 0 \\ 0 & \text{for } x = 0 \end{cases}$$

(a) Prove that f is not continuous at $x = 0$.
(b) Make a sketch of the graph of f.

****18.** (a) Prove that the function f given by $f(x) = x^2 - 2x + 3$ is continuous at $x = 4$.
(b) Show that f is continuous everywhere.
(c) Make a sketch of the graph of f.

****19.** A function f is defined by:

$$f(x) = \begin{cases} \frac{x^2 - 4}{x - 2} & \text{for } x \neq 2 \\ 5 & \text{for } x = 2 \end{cases}$$

(a) Decide whether f is continuous or not at $x = 2$.
(b) Make a sketch of the graph of f.
(c) If it is not continuous at $x = 2$, redefine f(2) so that f becomes continuous at $x = 2$.

****20.** A gun fires a shell at an initial velocity of 100 m/sec, at an angle θ to the horizontal. The gunner can determine θ. The target is 500 m away across level ground.
(a) What should the angle θ be?

(b) If the target is 2 m across, how accurately does the gunner need to control θ?
(In this question, take the only acceleration of the shell to be that due to gravity, 9.8 m/sec^2 downwards.)

(c) Why do so many questions in applied mathematics involve weapons?

****21.** Prove that the function f(x) defined by,

$$f(x) = \begin{cases} 0 & \text{if } x \text{ is rational} \\ 1 & \text{if } x \text{ is irrational} \end{cases}$$

is not continuous at any point.

****22.** If a function f(x) is defined in the neighbourhood of a point $x = a$ (but not necessarily at $x = a$), we say,

$$\lim_{x \to a} f(x) = b$$

if for all $\epsilon > 0$ there exists $\delta > 0$ such that if $0 < |x - a| < \delta$ then $|f(x) - b| < \epsilon$.

(a) Show that:

$$\lim_{x \to 2} \frac{1}{x} = \frac{1}{2}$$

(b) Show that a function f is continuous at $x = a$ if and only if,

$$\lim_{x \to a} f(x) = f(a)$$

(c) Show that $f(x) = 1/x$ has no limit at $x = 0$.

Calculating

***23.** A computer program calculates, for any input number x, the numbers,

$$\frac{x^1}{1!}, \frac{x^2}{2!}, \frac{x^3}{3!}, \ldots, \frac{x^n}{n!}$$

until the last number calculated, $\frac{x^n}{n!}$, is between ± 0.0001 (so, different x will require different n). For what values of x will the program need to calculate more than six of the terms?

****24.** A series for sin x is,

$$\sin x = x - \frac{x^3}{3!} + \frac{x^5}{5!} - \frac{x^7}{7!} + \ldots$$

(a) Show that because the terms are alternating in sign, the error after n terms (that is, $|\sin x - (\text{sum of the first n terms in the series})|$) is less than the absolute value of the last term taken.

(b) Suppose a calculator uses this series to calculate sin 5 to four decimal places. Find a bound on the number of terms it will need to use.

Series

***25.** Does $\frac{1}{2} + \frac{1}{4} + \frac{1}{6} + \frac{1}{8} + \ldots + \frac{1}{2n} + \ldots$ diverge? Prove your answer.

***26.** Use the ideas in the proof of the divergence of the harmonic series to find a number N such that:

$$1 + \frac{1}{2} + \frac{1}{3} + \ldots + \frac{1}{N} \geqslant 3$$

Is this the smallest such number? (Check with a calculator.)

****27.** Complete the proof that for $\epsilon > 0$ there exists N such that for all $n > N$,

$$1 > \frac{1}{2} + \frac{1}{4} + \frac{1}{8} + \ldots + \frac{1}{2^n} > 1 - \epsilon$$

*****28.** (a) According to Cauchy's condensation test for convergence: If $a_1, a_2, a_3, \ldots$ is a sequence such that:

$$a_1 > a_2 > a_3 \ldots$$

and,

$$\lim_{n \to \infty} a_n = 0$$

and,

$$a_1 + a_2 + a_3 + \ldots$$

converges, then,

$$2a_2 + 4a_4 + 8a_8 + \ldots + 2^n a_{2^n} + \ldots$$

also converges.

Give a proof of this.

(b) Hence show that,

$$\frac{1}{2 \log_e 2} + \frac{1}{3 \log_e 3} + \ldots + \frac{1}{n \log_e n} + \ldots$$

is divergent.

11
Counting

To show that the number of primes between 10 and 20 is four, there is nothing to do except list the primes between 10 and 20 and count how many there are. This could be called *proof by enumeration.* If a proof of some more general fact about counting is sought, it is often possible to use the idea that two sets are the same size if and only if their elements can be exactly paired or "put in one to one correspondence".

Example 1

Prove that a set of n elements has 2^n subsets.

(Here the empty set and the whole set are counted as subsets.)

Proof

It is possible to use mathematical induction, but it is more straightforward and leads to a better understanding, to pair off the subsets of the set of size n with some set that obviously has 2^n members. A set that obviously has 2^n members is the set of strings of 0s and 1s of length n, since there are two choices for the first place, 0 or 1, two for the second, and so on, so $2 \times 2 \times \ldots \times 2$ (n times), or 2^n choices, altogether. The final proof is:

Let the set of n elements be $\{a_1, \ldots, a_n\}$.

Let each subset be paired off with a string of 0s and 1s of length n by the rule: If a_i is in the subset, put 1 in the i'th place of the string (and if not, put 0).

Each different subset corresponds to a different string, (since if two subsets are different, some a_i is in one but not the other, so their corresponding strings differ in the i'th place).

Also, every string clearly corresponds to some subset.

Therefore, the number of subsets of a set of n elements is the same as the number of strings of 0s and 1s of length n, which is 2^n. ▮

Note

Since this proof is at quite a high level of abstraction, it would be helpful to give an explanatory example such as: If $n = 3$, the subset $\{a_1, a_3\}$ would correspond to 101 since a_1 is in the subset, a_2 is not, and a_3 is. However, the normal practice is to leave helpful examples out of the final versions of written proofs (in order, it often seems, to make them as difficult to read as possible).

Example 2

The number of r-element subsets of a set of n elements is denoted $\binom{n}{r}$. (This is sometimes described as, "the number of ways of choosing r things from n things"; however, "way of choosing" simply means "subset".)

Prove that,

$$\binom{n}{r} = \binom{n}{n - r}$$

The result says that the number of r-element subsets equals the number of $(n - r)$-element subsets. So we should pair off the r-element subsets with the $(n - r)$-element subsets. Now $n - r$ is the number that gives n when added to r, which suggests that, for each r-element subset, we should look at the *rest* of the whole set, that is, we should look at the "complement"—the set consisting of the remaining $n - r$ elements.

Proof

Pair off each r-element subset with its complement, (which is an $(n - r)$-element subset).

Clearly each different subset has a different complement, and each $(n - r)$-element subset is the complement of an r-element subset.

Therefore,

$$\binom{n}{r} = \binom{n}{n - r}$$

Notes

1. There is a formula for $\binom{n}{r}$ namely,

$$\binom{n}{r} = \frac{n!}{r!(n - r)!}$$

   We do not use this in the above proof because we have not proved that it is true. The proof of this formula is harder than the proof above—see Exercise 8 in this chapter.
2. As in the proof for Example 1, it is necessary to check that the correspondence really is a *pairing* of the two sets, that is, that different things in the first set are paired with different things in the second, and that everything in the second set is paired with something in the first. These two properties are sometimes expressed by saying that the correspondence is "one to one" and "onto". (See Chapter 12.)

Counting has important applications in computing. Computers can do wonderful things, as everyone knows, but it is sometimes surprising what they cannot do. One of the most significant limitations on their power is the phenomenon known as "combinatorial explosion". A computer often solves a problem by searching through a space of possibilities; for example, it might find the shortest round-trip route through a number of cities by checking the length of all round-trip routes through those cities. It is typical of such problems that the size of the space of possibilities grows very rapidly or "explodes", with the size of the problem, so that such "blind search" methods are restricted to quite small problems. Computers cannot, for example, play chess by blind search methods, since the space of possibilities to be searched is far too large. To know whether a proposed computer method of solving a problem is feasible, it is necessary to *count*, at least approximately, the number of steps the computer would have to perform.

A very important method of counting is to pair off the set to be counted with a set which is easy to count. This was illustrated in Examples 1 and 2, with further applications of the idea below.

Example 3

A backgammon competition starts with 32 players. In the first round they are matched in pairs. The winner of each game goes to the second round, and so on until there is one final winner. Show that the total number of matches played is 31.

Method 1

The most obvious, but not the best, method of counting the games is to count the number in each round. In the first round there are 16 games, in the second 8, in the third 4, in the fourth 2, and in the final round 1. So the total is $16 + 8 + 4 + 2 + 1 = 31$. ▌

Method 2

A better method is to note that the games can be paired off with the losers. For every game there is one loser, and the loser of each game is different, since losers drop out. But every player except the final winner is a loser. So the number of games is $32 - 1 = 31$. ▌

The superiority of the second method becomes even clearer when we apply it to games where the number of initial players is not a power of two, so that in some rounds there are byes. (See Exercise 4 at the end of this chapter.)

Example 4

Find the number of solutions of the equation,

$$x + y + z = 4$$

where x, y, z are non-negative integers, and prove that your answer is correct.

Method 1

We could list all the solutions by some systematic method, either by hand or by computer. The answer is 15. This is adequate here, but inefficient if the problem is larger, and in any case is not very illuminating.

Method 2

We prove that there is a pairing between the solutions of $x + y + z = 4$ and the 4-element subsets of a set of 6 objects. The 6 objects are thought of as 4 (identical) balls and 2 dividers. For each 4-element subset of {1, 2, 3, 4, 5, 6}, we place these 6 objects in a line, putting the balls in the places of elements in the subset (and the dividers in the remaining two places). By counting the objects in each compartment, we reach a solution of $x + y + z = 4$ — x is the number of objects in the first compartment, y is the number in the second and z is the number in the third. (For example, the subset {2, 3, 5, 6} gives the arrangement,

| . . | . .

and hence the solution $x = 0$, $y = z = 2$.) Clearly each solution of $x + y + z = 4$ comes from some 4-element subset of {1, 2, 3, 4, 5, 6}. Further, every subset gives a different solution, since we can reverse the process — given a solution of $x + y + x = 4$, we can find just one arrangement of ball and dividers, and hence one 4-element subset of {1, 2, 3, 4, 5, 6}, from which it comes.

So the number of non-negative integer solutions of $x + y + z = 4$ is equal to the number of 4-element subsets of a set of 6 objects, which is, $\binom{6}{4} = 15$. This uses the formula,

$$\binom{n}{r} = \frac{n!}{r!(n - r)!}$$

which is well-known, but ideally should be proved earlier. ▌

It is clear that this proof generalises to give a proof of the result: The number of non-negative integer solutions of,

$$x_1 + x_2 + \ldots + x_r = n$$

is,

$$\binom{n + r - 1}{n}$$

Example 5

The concept of "pairing off" is especially important for infinite sets. It was realised, at least as long ago as Galileo (1564–1642), that an infinite set, unlike a finite set, could be paired off with a subset of itself (other than the whole set itself). For example, the set of natural numbers can be paired off with its subset, the set of even numbers:

$$\begin{array}{cccccccccc} 1 & 2 & 3 & 4 & 5 & 6 & 7 & \ldots & n & \ldots \\ \updownarrow & \updownarrow & \updownarrow & \updownarrow & \updownarrow & \updownarrow & \updownarrow & & \updownarrow & \\ 2 & 4 & 6 & 8 & 10 & 12 & 14 & \ldots & 2n & \ldots \end{array}$$

This creates some problems about the "size" of infinite sets, since there does not seem to be any reasonable sense in which the set of natural numbers is "bigger" than the set of even numbers. It would be natural to think, after seeing this example, that probably *any* two infinite sets can be paired off with each other. That this is not the case was proved by Georg Cantor about 1873. He proved: "A set cannot be paired off with the set of its subsets". This is obvious if the set is finite since, as we showed, if the set has n elements, the set of its subsets has 2^n elements, and $n \neq 2^n$. However, although it is not obvious, the result is true for infinite sets as well.

Proof

As the result to be proved is of the form "there does not exist a pairing such that . . .", there is not much choice but to use a *reductio ad absurdum* which begins by supposing there is such a pairing and then deduces a contradiction.

Suppose the set A could be paired with the set of its subsets. For any $a \in A$, denote by p(a) the subset of A that is paired with a.

Now, (this is the hardest step of the proof) consider the subset of A:

$$B = \{a \in A \mid a \text{ is not a member of } p(a)\}$$

The condition makes sense because a is an element of A and p(a) is a subset. Now, to show that the subset B cannot be paired with any element of A, suppose there did exist $b \in A$ such that $p(b) = B$ and ask, is $b \in B$?

If b *is* in B, then b satisfies the condition for being in B, that is, b is not a member of p(b). However, $p(b) = B$, so this says that b is *not* in B, contradicting the assumption that b *is* in B.

On the other hand, if b is *not* in B, b does not satisfy the condition for being in B, so b is in $p(b) = B$. This is again a contradiction.

So whether $b \in B$ or not, a contradiction results. Therefore, there cannot exist any $b \in B$ such that $p(b) = B$. That is, B is a subset which is not paired with any element of A.

Since any proposed pairing p will produce such a subset B which is left unpaired, there cannot exist any pairing between A and the set of its subsets. ▌

Note

There is a double use of proof by contradiction. The overall method of the proof is to assume that there exists a pairing p and deduce a contradiction. This method includes the procedure of defining the set B (whose definition includes p, so that this defines a different B for each proposed pairing p), and showing that a contradiction results.

Example 6

Cantor used the same idea to prove: "The real numbers cannot be paired off with the natural numbers." (i.e. there are too many real numbers to be written in a list, even an infinitely long list).

Proof

The real numbers can be thought of as infinite decimals. Suppose there were a pairing,

1 $a_1\ a_2\ a_3\ a_4\ a_5\ \ldots$
2 $b_1\ b_2\ b_3\ b_4\ \ldots$
3 $c_1\ c_2\ c_3\ c_4\ c_5\ \ldots$
4 $d_1\ d_2\ d_3\ d_4\ d_5\ d_6\ \ldots$
. .
. .
. .

(In each decimal, the decimal point is put in wherever it occurs.)

Then the real number $0.z_1\ z_2\ z_3\ z_4 \ldots$ where $z_1 \neq a_1$, $z_2 \neq b_2, \ldots$ is not paired with any integer. That is, we take the "diagonal" $a_1\ b_2\ c_3\ d_4 \ldots$ and change every digit; the resulting number is not in the list since it differs from every number in the list in at least one place.
So there is no pairing of the real numbers with the natural numbers. ▌

One thing should be clear from this chapter. Although counting is a part of mathematics which we begin learning even before we go to school, it does not follow that it is easy. Counting is hard.

Exercises

(Gradings of exercises: * easy, ** moderate, *** difficult.)

***1.** List the eight strings of 0s and 1s of length 3, and write next to each the subset of $\{a_1, a_2, a_3\}$ with which it is paired (according to the pairing in Example 1 above).

***2.** Comment on the reasoning: "The set of whole numbers is really bigger, in some sense, than the set of even whole numbers, since if one army general had a soldier for every whole number, and his opponent only had one for every even number, and they lined up facing one another, the first general would obviously win."

****3.** Find the number of solutions of the equation,

$$a + b + c + d = 11$$

where:
(a) a, b, c, d are non-negative integers.
(b) a, b, c, d are positive integers (that is, none of them are zero). Hint: Pair the solutions with certain solutions of,

$$a + b + c + d = 7$$

(c) $a \geqslant 1$, $b \geqslant 2$, $c \geqslant 0$, $d \geqslant 3$.

****4.** A backgammon competition starts with 24 players. In the first round they are matched in pairs. The winner of each game goes on to the second round and so on, until there is one final winner. If there is an odd number of players in a round, one of them goes to the next round without playing, while the rest play in pairs. Find the total number of matches played. Generalise to n players.

****5.** Prove that if in a sample (that is, a set) of white and black balls there are 51 per cent white balls (to the nearest 1 per cent) then the size of the sample is at least 35.

*****6.** Show that there is a pairing between the natural numbers and the rational numbers.

*****7.** Show that there is a pairing between the set of real numbers between 0 and 1 and the set of all real numbers.

*****8.** Show that, $\binom{n}{r} = \dfrac{n!}{r!(n-r)!}$.

12

Functions

The notion of "function" unifies many of the things studied in mathematics. Whatever can be thought of under the formula, "For anything of a certain kind, there is a thing of another kind" can be regarded as a function. For example, the temperature in a room may vary from point to point; we say that the temperature is a function of the position in the room. That is, for every point in the room, there is a real number which is the temperature at that point.

Whenever there is, for every point in a set A, an associated point in a set B, we say there is a function f from A to B. This is denoted $f: A \to B$. The possibility that $A = B$ is not excluded. A function is also called a *map* or a *transformation*. The element of B associated to $a \in A$ is written f(a) and is called the *image of a under f* or the *value of f at a*.

Examples

1. The temperature T in a room is a function T:{points in the room} → {real numbers}. In fact *any* quantity that varies is a function. Similarly, the density of a body is a function of position and the velocity of a body is a function of time.
2. Functions from **R** to **R** are the functions usually drawn on graphs. For example, the function $f: \mathbf{R} \to \mathbf{R}$ given by the formula $f(x) = x^2$ has the familiar graph shown in Figure 12.1.

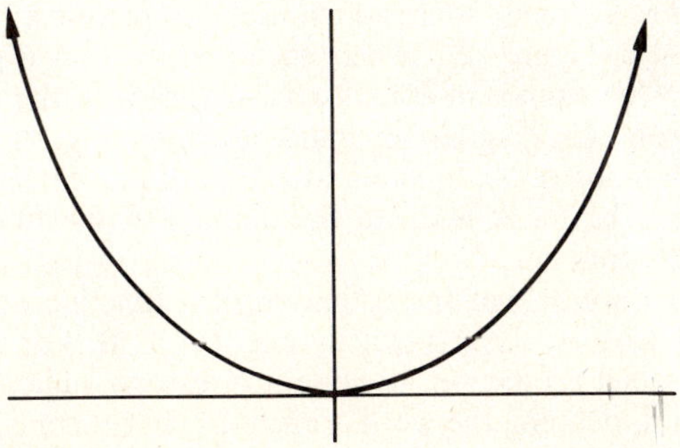

Figure 12.1

3. A typical computer program reads input, calculates with it, and produces an output. The program thus provides a function from the set of possible inputs to the set of possible outputs. Each input gives one output. (There is the possibility that a bad program might not give an output for some inputs; that is, it might calculate indefinitely. It would be better to say that *good* computer programs all define functions from inputs to outputs).
4. Any formula, for example $f(x, y) = x^2 - x^3y$, defines a function. In this case, f produces, for any pair of real numbers (x, y), a single number $x^2 - x^3y$, and so f is a function $\mathbf{R}^2 \to \mathbf{R}$. Thus, f is called a "function of two variables".
5. On a map with contours, the height of land above sea level is a function of position. This can be regarded as a function of two variables, since it takes two co-ordinates to specify a position, for example, "the height of land at 157° 27′ 2″ W, 34° 0′ 2″ S is 750 m".
6. A function $f: \mathbf{N} \to \mathbf{R}$ means that for each natural number n there is associated a real number f(n). That is, a function $f: \mathbf{N} \to \mathbf{R}$ is really the same thing as an infinite sequence of real numbers. In general, for any set X, a function $f: \mathbf{N} \to X$ is just a sequence of elements of X.
7. The velocity of a moving body is a "vector function of time", which means that at any given time the body has a (possibly different) velocity, which is a vector. A body which is initially at the origin, and whose velocity is given by $v(t) = (0, 0, t)$, moves along the z-axis with an increasing velocity.
8. If $A = \{a, b\}$ and $B = \{c, d\}$, there are four functions from A to B, namely, f_1, f_2, f_3, f_4 defined by,

$$f_1(a) = c, \quad f_2(a) = d, \quad f_3(a) = c, \quad f_4(a) = d$$
$$f_1(b) = d \quad f_2(b) = c \quad f_3(b) = c \quad f_4(b) = d$$

9. The examination results you get are a function of the amount of effort you put in (and of some other variables as well).

Two functions from A to B are defined to be equal when their values are equal for all $a \in A$. That is, if $f: A \to B$ and $g: A \to B$ are two functions,

$$f = g \text{ iff } f(a) = g(a) \text{ for all } a \in A$$

Thus the functions, $f: \{0, 1\} \to \{0, 1\}$ given by $f(x) = x^2$ and $g: \{0, 1\} \to \{0, 1\}$ given by $g(x) = x$, are equal, though the formulas are different. Two functions between different sets are never called equal. Thus $f = \mathbf{R} \to \mathbf{R}$ given by $f(x) = x^2$ and $g = \mathbf{Z} \to \mathbf{Z}$ given by $g(x) = x^2$ are different functions, though their formula is the same.

There is no thinking without an image, as the ancient sages say. Just as it is helpful to picture sets by Venn diagrams, so it is often useful to picture a function $f: A \to B$ by a diagram such as that in Figure 12.2, where each point $a \in A$ is joined by an arrow to the point $f(a) \in B$. (One cannot of course actually draw all the arrows if the sets are large or infinite.) With this diagram in mind we often say, "f takes a to b" when $f(a) = b$ — we are thinking of the function as a kind of movement of the elements of A to the elements of B. The set A is called the domain of the function f. The set B is sometimes called the codomain.

It is important to notice that the definition of a function $f: A \to B$ includes the notion that for every $a \in A$, there is exactly one $f(a) \in B$. For example, the formula $f(x) = \pm\sqrt{x}$ is not called a function, because it gives two values for each x. This is a matter of convention in defining the word "function", but there is a good reason for it.

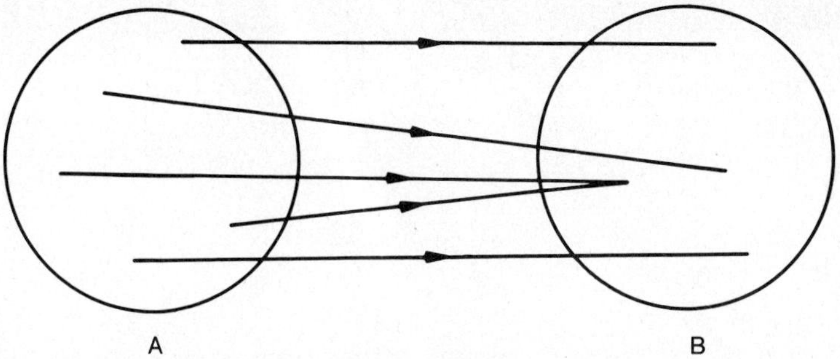

Figure 12.2

The cases that arise most commonly are of the kind where for every input, there is only one possible output; a formula gives just one answer when a number is put in it; there is only one temperature at any place, and so on.

The other important point to notice about the definition of a function is that, though every $a \in A$ is taken to something by f, it is not required that every $b \in B$ should come from something in A. Thus a typical function has a diagram such as that in Figure 12.3, where the part of B above the dotted line contains the elements of B which are not f(a) for any $a \in A$. Again this is a convention of the way "function" is defined, but again there is a good reason for it. Typically it is obvious what kind of things f applies to (the set A), and the general kind of thing the answers f(a) should be (the set B), but it is not obvious which of those actually appear. For example it is obvious that the temperatures in a room are numbers, but before measurements are made it is unknown which numbers will appear.

It is natural however to consider that part of B which does consist of the values taken by the function. This is called the *range* or *image* of f. In symbols, range of $f = \{b \in B: b = f(a) \text{ for some } a \in A\}$. Remember that as usual "some" means "at least one", not "exactly one". (Note also that the use of the word "image" here is different from its use in the phrase "image of a under f".) For example, the range of the function $f: \mathbf{R} \to \mathbf{R}$ given by,

$$f(x) = x^2$$

is,

$$\{x \in \mathbf{R} \mid x \geqslant 0\}$$

It can happen that the range of f is the whole of B. In that case f is called *onto* (or *surjective*), since the f(a)'s "cover" the whole of B. That is,

f is *onto* if for every $b \in B$ there is an $a \in A$ such that $f(a) = b$

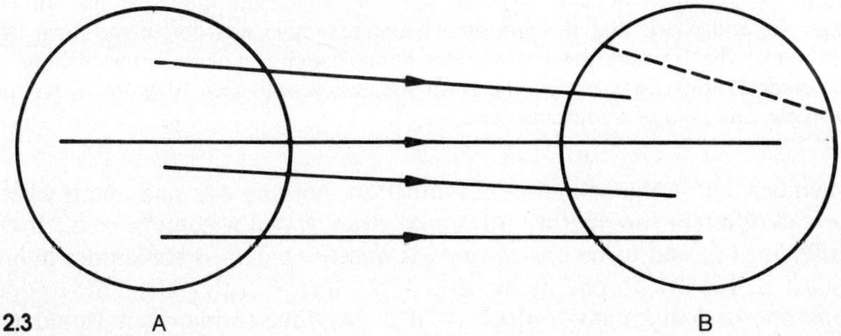

Figure 12.3

Example 1

Is $f: \mathbf{R}^2 \to \mathbf{R}$ given by $f(x, y) = x + y$ onto? Prove your answer.

Answer

Yes.

Let $z \in \mathbf{R}$.

Then, $z = f(z, 0)$.

So, every $z \in \mathbf{R}$ is of the form $f(x, y)$ for some $x, y \in \mathbf{R}$.

So, f is onto. ▌

Note

Since the statement to be proved was of the form, "For all $z \in \mathbf{R}$, there is . . .", it was necessary to begin with "Let $z \in \mathbf{R}$" (as happens with any "all" statement). Then to prove the "there is", it was necessary to find, for any $z \in \mathbf{R}$, one thing satisfying the condition. As usual, when a statement to be proved involves "there is", only one of the many things satisfying the condition is exhibited: we wrote $z = f(z, 0)$, but we could equally well have written,

$$z = f\left(\frac{z}{2}, \frac{z}{2}\right)$$

Example 2

Is $f: \mathbf{R} \to \mathbf{R}^2$ given by $f(t) = (t, t^2)$ onto? Prove your answer.

Answer

No.

$(2, 1)$ is not of the form $f(t)$ for any t in $\mathbf{R}$, since $1 \neq 2^2$.

So f is not onto. ▌

Note

In this example, because we had to prove an "all" statement false, we had to exhibit a counterexample, and check that the proposed counterexample was not of the form f(t). If, in another example, the formula for f was more complicated and it was not obvious that the proposed counterexample was not of the form f(t), we would need to write f(a) = proposed counterexample, and deduce a contradiction.

Remember that in the definition of a function, nothing was said about whether the f(a)'s were all different, or whether f(a) could equal f(a′) for some $a \neq a'$. If the f(a)s are all different, f is said to be *one-to-one* (or *injective*). This is sometimes abbreviated as 1–1. That is, f is one-to-one if, for all $a \neq a'$, $f(a) \neq f(a')$ (thus "one-to-one" is to be read as opposed to "many-to-one" — if f was "one-to-many" it would not be a

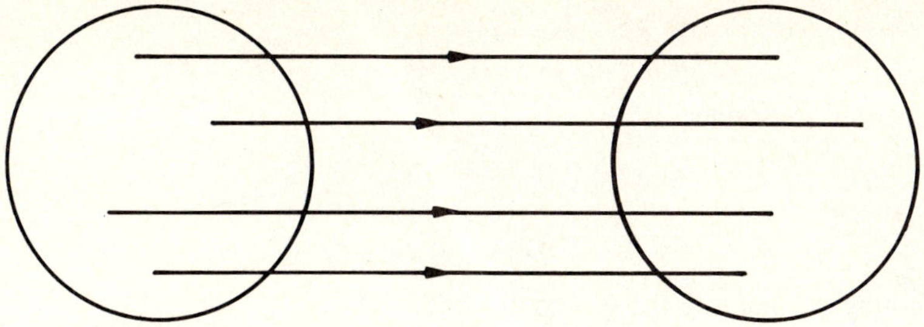

Figure 12.4

A B

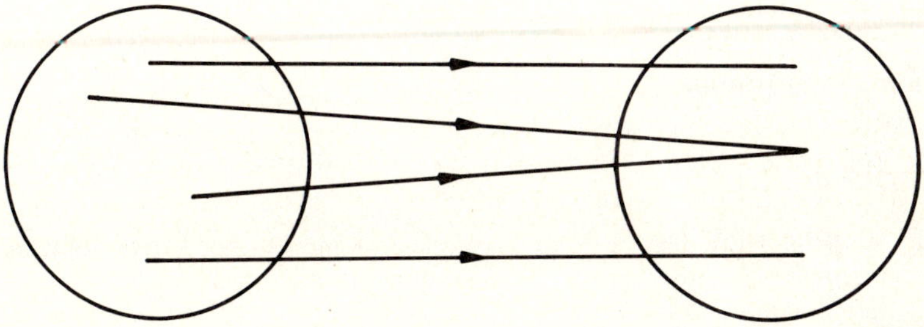

Figure 12.5

function at all. That is, f is one-to-one if it doesn't take any two things (in A) to the same answer (in B). This is shown in Figure 12.4. Figure 12.5 is an example of what *cannot* happen if f is one-to-one. Being one-to-one excludes a case like this: If there are two distinct a, a′ anywhere in A such that $f(a) = f(a')$, then f is not one-to-one.

When proving that a given function is one-to-one, it is usually best to use the contrapositive of the definition:

$$f \text{ is one-to-one if, whenever } f(a) = f(a') \text{ then } a = a'$$

This can be loosely expressed in words as, "If f takes two elements of A to the same answer, then the two elements must have been the same to begin with". Recall that the contrapositive of a statement is logically equivalent to it. (See Chapter 6.)

Example 3

Is the function $f: \mathbf{R}^2 \rightarrow \mathbf{R}^3$ given by, $f(x, y) = (x, x + y, 2x - y)$ one-to-one? Prove your answer.

Answer

Yes.

Let,

$$f(x, y) = f(x', y')$$

That is,

$$(x, x + y, 2x - y) = (x', x' + y', 2x' - y')$$

So, equating co-ordinates,

$$x = x'$$
$$x + y = x' + y'$$
$$2x - y = 2x' - y'$$

Substituting the first into the second, we find,

$$x' + y = x' + y'$$

so,

$$y = y'$$

So $x = x'$ and $y = y'$, so $(x, y) = (x', y')$.

Therefore f is one-to-one.

Example 4

Is $f: \mathbf{R}^2 \to \mathbf{R}^2$ given by $f(x, y) = (x - y, y - x)$ one-to-one? Prove your answer.

Answer

No.

As $f(1, 1) = f(2, 2)$, f is not one-to-one. (As always, an "all" statement is disproved by giving one counterexample.)

Note

Since the method of proof depends on whether the answer is yes or no to the questions "Is f onto?". "Is f one-to-one?", it is necessary to know the answer before writing down the proof. To discover what the answers are, an understanding of the function is needed. If one has no understanding of the function to begin with, calculating a few values may help. An attempted proof that f *is* onto or one-to-one, may also help. If it works, the question is finished. If not, it may give enough insight to show where a counterexample may be found.

A function may be both onto and one-to-one. In this case it is called *bijective* (or a *bijection*, a *one-to-one correspondence* or a pairing). It might be expected that combining the notions of onto and one-to-one would produce a more difficult notion, but in a way the opposite is true. A bijection $f: A \to B$ is just a *pairing* of the elements of A with the elements of B. Every element of A is paired with exactly one element of B (because f is one-to-one) and every element of B is used (because f is onto). Pairing of two sets is a very natural idea (every footballer has to cover his opposite number, a waiter puts down a fork with every knife) and only when it is necessary to *prove* that two sets are paired, is it realised that pairing really involves two separate concepts, being one-to-one and being onto.

Example 5

Let A be a set with n elements and let B be the set of all (n − 1)-element subsets of A, for example, if A = {a, b, c}, B = {{a, b}, {a, c}, {b, c}}.

Let f: A → B be defined by f(a) = A − {a}, that is, A with a left out.

Prove that f is a bijection.

Proof

Let,

$$f(a) = f(a')$$

So,

$$A - \{a\} = A - \{a'\}$$

Then a must equal a′, since otherwise a′ would be in the left hand side but not in the right hand side.

So,

$$a = a'$$

That is, f is one-to-one.

Now, let,

$$X \in B$$

That is, X is an n − 1 element subset of A. Since A has n elements, there is one element of A that is left out of X. Call it x.

Then,

$$X = A - \{x\}$$

So,

$$X = f(x)$$

So, X is the image of some $x \in A$.

That is, f is onto.

So, f is bijective, since it is one-to-one and onto. ▌

Exercises

(Gradings of exercises: * easy, ** moderate, *** difficult.)

*1. (a) What is meant by an onto (or surjective) function?
 (b) Is the function f: **R** → **R** defined by f(x) = sin x, onto? Prove your answer.
 (c) Is the function f: **R** → [−1, 1] defined by f(x) = sin x, onto? Prove your answer.

(d) A function g is onto. What can you say about its range and codomain?

*2. (a) Is the function $f: \mathbf{R}^2 \rightarrow \mathbf{R}$ given by $f(x, y) = 3x - y$, onto? Prove your answer.
(b) Is the function $f: \mathbf{R} \rightarrow \mathbf{R}^2$ given by $f(t) = (1 + t, t + t^2)$, onto? Prove your answer.

*3. (a) What is meant by a one-to-one (or injective) function?
(b) Is the function $f: \mathbf{R} \rightarrow \mathbf{R}$ defined by $f(x) = x^2$, one-to-one?
(c) Is the function $f: \mathbf{R} \rightarrow \mathbf{R}$ defined by $f(x) = x^3$, one-to-one?

*4. (a) Is the function $f: \mathbf{R}^2 \rightarrow \mathbf{R}^3$ given by $f(x, y) = (x + y, x, y)$, one-to-one? Prove your answer.
(b) Is the function $f: \mathbf{R}^2 \rightarrow \mathbf{R}^2$ given by $f(x, y) = (1, x + y)$, one-to-one? Prove your answer.

*5. Show that the function $f: \mathbf{R}^2 \rightarrow \mathbf{R}^3$ given by $f(x, y) = (x + y - z, x + y, y - z)$, is not onto.

*6. (a) What is meant by a bijective function (or a bijection)?
(b) Is the function $f: \mathbf{R} \rightarrow \mathbf{R}$ defined by $f(x) = x^3$, a bijective function?
(c) Is the function $f: \mathbf{R} \rightarrow \mathbf{R}$ defined by $f(x) = x^2$, a bijective function?

*7. (a) Is the function $f: \mathbf{R} \rightarrow \mathbf{R}$ defined by $f(x) = e^x$ bijective?
(b) Is the function $f: \mathbf{R} \rightarrow (0, \infty)$ defined by $f(x) = e^x$ bijective?

*8. Show that the function $f: \mathbf{Z} \rightarrow \mathbf{Z}$ given by $f(z) = -z$, is a bijection.

**9. (a) Prove that if A is a set with n elements and B is a set with m elements, then the number of different functions from A to B is m^n. (If you cannot start, try some small examples, say with $A = \{a, b\}$, $B = \{X, Y, Z\}$)
(b) If $n < m$, show that the number of 1–1 functions from A to B is $m(m - 1) \ldots (m - n + 1)$
(c) If $m = n$ show that the number of different bijections from A to B is $n!$.

**10. The function F: {functions $\mathbf{R} \rightarrow \mathbf{R}$} $\rightarrow \mathbf{R}$ is given by $F(f) = f(0)$. (F is called "evaluation at 0", which is an obvious name once you understand what F does). Is F onto? Prove your answer.

***11. Show that for any set A the function F: {functions $A \rightarrow \{0, 1\}$} $\rightarrow$ {subsets of A} given by $F(f) = \{x \in A: f(x) = 1\}$ is a bijection.

13

Real-life problems

"Mathematics trains you to think logically." This proposition is widely believed, and consequently mathematicians have received large sums of money to force their product down the throats of generations of reluctant students. But is it true? If it is, what does mathematics teach you to think logically *about*? On the face of it, learning algorithms to do long division, solve linear equations or evaluate complicated integrals does not seem very relevant to thinking about problems in real life. Indeed, employers continually complain that graduates know methods and techniques, but not when to apply them, because they have not learned to think logically about real problems.

Previous chapters illustrated the importance of logic in *analysing* and making precise various abstract concepts. For example, in Chapter 4, "a divides b" was analysed as "There exists c such that b = ac", and in Chapter 10 the notions of limit, convergence and continuity were analysed using "all" and "some". Logic in this sense is certainly relevant in dealing with real-life problems, where one must analyse a situation logically to find those features which are important for making a prediction or plan.

The related topics of compound interest and exponential growth and decay are easy mathematically but illustrate many of the ideas used in analysing a real situation and making a mathematical model of it.

Example 1 (Compound interest)

A building society asks in an advertisement, "Are you getting the little 10% or the big 10%?" By "the little 10%" they mean that an investor has 10 per cent of his investment added at the end of each year. By "the big 10%" they mean 10 per cent annual interest, compounded daily—that is, 10/365 per cent added each day. What annual rate of interest is "the big 10%" equivalent to?

Solution

Of course there is a compound interest formula which can be looked up, but the point here is to learn how to analyse such situations, so that one can deal with more complicated cases oneself. One of the difficulties with "formulas" and "methods" is that they never quite seem to cover the situations one actually meets. There is always a temptation to use the formulas outside their limits of applicability.

The information that, "10/365% is added each day" needs to be expressed mathematically. This will give the relationship that holds between the amount in the investment on one day and the amount on the next.

Firstly we set up symbols for these amounts: Let x_i be the amount in the investment on the i'th day.

Then for each i = 0, 1, . . . , 364,

$$\begin{aligned} x_{i+1} &= x_i + \frac{10}{36500} x_i \\ &= \left(1 + \frac{10}{36500}\right) x_i \end{aligned}$$

We are asked about the amount at the end of the year, compared to how much there was to begin with, that is, the relation between x_{365} and x_0.

Now,

$$x_1 = \left(1 + \frac{10}{36500}\right) x_0$$

and,

$$\begin{aligned} x_2 &= \left(1 + \frac{10}{36500}\right) x_1 \\ &= \left(1 + \frac{10}{36500}\right)^2 x_0 \end{aligned}$$

and so on, so that for any i,

$$x_i = \left(1 + \frac{10}{36500}\right)^i x_0$$

(We have used a simple mathematical induction argument.)

So, putting i = 365,

$$\begin{aligned} x_{365} &= \left(1 + \frac{10}{36500}\right)^{365} x_0 \\ &= 1.105 x_0 \end{aligned}$$

using a calculator. So the effective annual rate of interest is 10.5 per cent; this is not a great deal more than 10 per cent, but enough to be worth considering.

Notes

These features of the example tend to occur often.

1. A mathematical equation was found expressing the "local structure" of the situation; in this case,

$$x_{i+1} = \left(1 + \frac{10}{36500}\right) x_i$$

expresses the relation between the amounts in the investment on *neighbouring* days. Such an equation is of course an "all" statement; the relation is the same for all pairs of consecutive days.

2. In a quite separate step, this equation is "solved" to find the "global" structure resulting from the "local" structure expressed in the equation. In this case we want to know how the daily additions of interest "add up" over the whole year, to give the relation between the amounts at the beginning and end of the year.

The next example is a continuous situation, unlike the previous one, which was discrete. Consequently, it needs some simple techniques from calculus, but otherwise it is very similar.

Example 2 (Exponential decay)

A tank of 100 litres of water has 30 grams of salt dissolved in it. Fresh water enters it at the rate of 4 litres per minute. This is mixed well and the tank overflows, also at the rate of 4 litres per minute. How much salt is there in the tank at any later time?

Solution

It is clear that the amount of salt should gradually decrease from 30 grams, but never reach zero.

We let $x(t)$ grams be the amount of salt in the tank at any time $t > 0$. (Thus $x(0) = 30$, and $x(t)$ is a function of time which we have to determine.)

Next we consider what information we have about $x(t)$. We are given something about how $x(t)$ varies, that is, about dx/dt. The rate at which $x(t)$ is increasing (at any time) is dx/dt; its units are grams per minute (units of x per units of t). How many grams per minute of salt are coming in and going out? None are coming in (since the inflow is fresh water), but 4/100 *of the salt in the tank* is flowing out per minute (because 4/100 of the solution is flowing out, and the salt is well mixed). That is,

$$\frac{dx}{dt} = -\frac{4}{100}x$$

This differential equation, together with the initial condition $x(0) = 30$, expresses all the information in the problem.

This equation must now be solved to find x (in terms of t). One method is to invert both sides and integrate with respect to x; another is separation of variables.

The solution is,

$$x(t) = 30e^{-\frac{4}{100}t}$$

Notes

1. As in the previous example, the problem was to convert the given local information (here, the rate of outflow at any time) into symbols, so that the resulting equation could be solved to give the global answer (a formula for the amount of salt after any time).
2. As with many cases, especially continuous ones, the model is only an approximation to the real situation. The solution of salt will not really be mixed perfectly, and at the atomic level the situation is not continuous at all, but discrete. One needs some reason to believe that these approximations will make the model and its prediction only a little inaccurate. In this

example, a small change in the rate of inflow or the initial condition would make only a small change in x(t).

The related concepts of *feedback* and *stability* provide examples of how mathematical concepts are used in understanding the world. On the night of 27/28 July 1943, the Royal Air Force bombed Hamburg with an unusually large force of 787 planes. Photographs taken later revealed that something out of the ordinary had happened. Small fires created by the bombing usually burned separately and were easily controlled, but this time were enough to create an updraught effect. As the hot air rose, it drew a wind into the city, fanning the flames and creating an even stronger updraught and winds of up to 150 km/h. This *feedback* effect, where the fires created the conditions for even more fire, caused a firestorm which burnt out much of the city; 40,000 people died. The city was visited by Dr Goebbels, Minister for Public Enlightenment and Propaganda, who consoled the survivors with the words, "Our walls are broken, our hearts are not". R.A.F. Bomber Command tried hard to repeat their success (or war crime, depending on one's point of view) in Berlin, but never succeeded; some other cities suffered partially similar fates.

Such phenomena, where small changes die out but large changes result in *instability* because of feedback, arise in many important cases, but are very hard to analyse. What level of economic activity would cause an underdeveloped country's economy to "take off" and become self-sustaining? How much atmospheric dust from a nuclear war or large volcanic eruption would cause major glaciation, so that the earth would reflect more heat and enter an ice age? What level of radioactivity in a nuclear reactor core would cause an uncontrollable chain reaction and meltdown?

There are also cases of *negative feedback*, where a process restricts itself rather than feeds itself as in the examples above. Such a process is found in the temperature control system of the body, which initiates sweating or shivering when the temperature of the body diverges from what is desirable. The following example is one of negative feedback.

Example 3 (Falling bodies)

Do heavier bodies fall faster? Investigate.

Answer

Of course they do, at least sometimes. Rocks fall faster than feathers. But it is well-known that heavy bodies (in air) all fall (initially) with the same constant acceleration. Galileo discovered this, or at least asserted it. He is supposed to have dropped two weights from the Leaning Tower of Pisa and found that they hit the ground at the same time (though in fact a whole book has been written to show that there is no evidence that Galileo ever performed such an experiment, either in Pisa or anywhere else). Such an experiment does not, in any case, show that the acceleration of falling bodies is constant, and indeed this fact is quite hard to observe without sophisticated techniques such as slow-motion photography, since the rate of fall is so fast. When using primitive equipment, such as Galileo and earlier scientists had, it is necessary to slow down the fall to observe it. One method might be to roll bodies down an inclined plane; Galileo did this. Another method might be to use light bodies, or to let heavy ones fall in a medium in which they fall more slowly, such as water. The two methods yield different answers. A heavy body on an inclined plane falls with (approximately) constant acceleration (and hence increasing velocity), while in the other cases, bodies fall with constant *velocities*, these velocities being approximately proportional to their weights. The difference is now explained by the *resistance* of the air or water, though it was certainly not always so obvious.

How can we incorporate a treatment of resistance in a mathematical analysis? Is the proportionality of speeds to weights in a resisting medium easily explained? Without resistance, the acceleration of a falling body would be a constant $g = 9.8 \text{ m/sec}^2$, that is,

$$\frac{dv}{dt} = g$$

where v is the velocity at any time. What should the resistance be? Plainly it should increase with v, and be zero when v is zero. The simplest assumption is that it should be proportional to v. Is this reasonable? It could be tested experimentally, or it could be argued for in this way: At twice a certain speed, twice as many air particles have to be pushed out of the way per unit time, so they would have twice the retarding effect on the body. This argument is not totally convincing, and experiments, for example in a wind tunnel, should be performed to check. As it happens, this is approximately correct for quite a wide range of values of v, though we should, as always, be clear about what range of values our "all" statement is true for. Thus the acceleration due to resistance is also proportional to v, and hence,

$$\frac{dv}{dt} = g - kv$$

where k is a constant. This is a first order linear equation (in the unknown function of time v). It can be solved using the integrating factor e^{kt},

$$\frac{dv}{dt} + kv = g$$

$$e^{kt}\frac{dv}{dt} + k\,e^{kt}\,v = g\,e^{kt}$$

$$\frac{d}{dt}(e^{kt}\,v) = g\,e^{kt}$$

$$e^{kt}\,v = \frac{g}{k}e^{kt} + C$$

If the body is dropped from rest, $v = 0$ at $t = 0$, so $C = -\frac{g}{k}$.

Thus,

$$e^{kt}\,v = \frac{g}{k}(e^{kt} - 1)$$

$$v = \frac{g}{k}(1 - e^{-kt})$$

So as $t \to \infty$, $v \to \frac{g}{k}$, the "terminal velocity".

How is this result affected by differences in the mass of the body? *For any particular body*, k was a constant (i.e. the differential equation is an "all" statement about all velocities and all times, but not about all bodies; k is independent of t and v, but depends on m). However, the argument above allows us to see what effect different masses should have, that is, it allows us to generalise to an "all" statement about all masses. How does k vary as m varies? That is, what function of m is k? Consider the resistance for a fixed v but varying m. The force of resistance is not affected by m (how the air particles are pushed out of the way depends on the size, shape and velocity of the body, but not on its mass). Thus, in the equation,

$$\text{force} = \text{mass} \times \text{acceleration}$$

the force is constant (for fixed v), so the acceleration is proportional to $1/m$. So k can be replaced by K/m, where K is now a constant for all velocities and times and all masses (within reasonable ranges).

So the terminal velocity is now $g/(K/m)$, which is gm/K. Thus, we find that the terminal velocity is proportional to the mass; this is in accordance with experiments.

Then what use is a parachute? A body with a parachute has a much lower terminal velocity than one of the same weight without a parachute. The above argument implicitly dealt with bodies of a fixed shape and size—the argument about pushing twice as many air particles out of the way would not be right otherwise. Obviously a larger surface area creates greater drag, but here we will not attempt a further level of generalisation to deal with this.

Further reading: T. L. Saaty and J. M. Alexander, *Thinking with Models* (1981), Pergamon, Oxford, UK.

Exercises

(Gradings of exercises: * easy, ** moderate, *** difficult.)

*1. A prisoner can earn one day of remission for every three days served. If the sentence is for ten years, when will the prisoner actually be released if he/she earns maximum remission?

*2. If a population grows at an (instantaneous) rate of 2 per cent a year (i.e. $\frac{dP}{dt} = 0.02P$), how long does it take to double?

**3. If a population grows according to,

$$\frac{dP}{dt} = 0.02P$$

and starts with $P = P_0$ at $t = 0$, will the value of P at $t = 1$ be $1.02P_0$? Explain.

**4. A rule often given for the growth of populations is this: "If a population grows at x per cent a year, it takes about $\frac{70}{x}$ years to double". Investigate this rule. Is it true for large x? Small x? Why 70?

**5. An investor puts $500,000 into Hitek Bonds, which pay a profit of 20 per cent a year, compounded daily (so if P is the amount of money in the bonds at any time, $\frac{dP}{dt} = 0.2P$ approximately).

(a) How much money would the investor own at the end of a year?

(b) A tax accountant advises that this is overprofitable, and suggests that the investor take out \$200 of the interest each week and put this towards the hire of a Volvo, which would be a tax write-off under the government's Small Business Incentive Scheme. The accountant also suggests that at the end of each six months, all the remaining interest should be taken out and invested in the Lake Eyre Ricegrowers' Cooperative, which loses money continuously at a rate of 10 per cent a year $\left(\frac{dP}{dt} = -0.1P\right)$. This would qualify the investor for various advantages under the Rural Rorts Scheme. If the investor takes this advice, what will the total of both investments be after one year? (Take it that there are 52 weeks in a year.)

****6.** (a) The solution to the equation,

$$x_{i+1} = A\,x_i$$

is obviously,

$$x_i = A^i x_0$$

since $x_1 = Ax_0$, $x_2 = Ax_1 = A^2x_0$, and so on. (Compound interest is a case in point, where A is 1 plus the rate of interest.) The equation,

$$x_{i+1} = A\,x_i + B$$

for constants A, B can be solved by looking for a solution of the form,

$$x_i = f(i)A^i$$

(i.e. a function of i times A^i, instead of a constant times A^i, as for $x_{i+1} = A\,x_i$). Find what f(i) must be, and prove the result by mathematical induction.

(b) A bank lends a customer \$50,000 at 15 per cent interest, calculated monthly (i.e. 1.25 per cent per month), with repayments of \$700 per month. How long does it take to pay off the loan? (Use the result obtained in part (a).)

(c) An aquarium initially contains 10 litres of fresh water (and some fish). The fish are fed once a day with food which contains .10 gram of salt (which the fish do not eat and which stays dissolved in the water). To prevent salt build-up, 2 litres of the water are taken out and replaced by fresh water at the end of every week. What is the eventual concentration of salt? (Use the result obtained in part (a).)

****7.** Two tanks, A and B, each contain 100 litres of water. Initially A contains 30 grams of salt while the water in B is fresh. Fresh water flows into A at 4 litres per minute, is mixed well in A, and overflows into B, also at 4 litres per minute. The water in B is mixed well and overflows at 4 litres per minute. When is the amount of salt in B a maximum? (This requires a method for solving first order linear differential equations.)

****8.** It is estimated that the population growth rate of a developing country will fall linearly from 2 per cent a year to 1.5 per cent a year over the next 10 years. If the present population is 20 million, what is the estimated population in 10 years? (This requires the separation of variables technique for solving differential equations.)

**9. An initially unpolluted lake of 10^9 litres has a river flowing through it at 100,000 litres a day. A factory is built which discharges 1,000 litres a day of pollutant into the lake.
   (a) What will be the eventual level of pollution in the lake?
   (b) How long will it take to reach half this level?
   (c) Is there anything unrealistic in the problem? (This requires a technique for solving first order linear differential equations.)

**10. A light hangs above the centre of a table of radius r metres. The illumination at any point on the table is proportional to the cosine of the angle of incidence and is inversely proportional to the square of the distance from the light. How far should the light be above the table to give the strongest illumination at the edge of the table? (If you are not sure which is the angle of incidence, draw a diagram and work out which it must be.) Do the proportionalities in the question seem reasonable to you, or do you think they are just made up? If they do seem reasonable, can you derive them from some simpler assumptions?

**11. The formula for the area of a circle (πr^2) is the integral of the formula for the circumference ($2\pi r$); the formula for the volume of a sphere ($\frac{4}{3}\pi r^3$) is the integral of the formula for the surface area ($4\pi r^2$). Explain why.

***12. In the eighteenth century, there were seven bridges across the River Pregel at Koenigsberg, as shown in Figure 13.1. Show that it is impossible to walk over all seven bridges once, without walking over one of them twice.

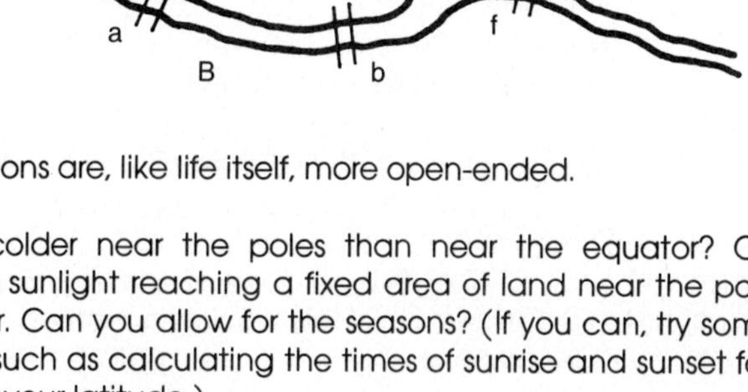

Figure 13.1

The following questions are, like life itself, more open-ended.

*13. Why is it colder near the poles than near the equator? Compare the amounts of sunlight reaching a fixed area of land near the poles and near the equator. Can you allow for the seasons? (If you can, try something more ambitious, such as calculating the times of sunrise and sunset for any day of the year, at your latitude.)

**14. Is it better to buy small fruit or large fruit? How much does the answer depend on the difference in price?

**15. Why do mice scurry but elephants walk slowly?

**16. Why do the large cornflakes rise to the top of the packet?

**17. You are asked to predict the spread through the population of a certain sexually-transmitted disease. What information do you need to know to make reasonable predictions?

**18. "If a string has an end, it has another end." Is this true? Is it true under some interpretations of "string" but not others? Would you expect to be able to prove it?

**19. (Laboratory session) Leave a glass of water to stand for half an hour or so, so that the water is still. Gently lower a drop of ink onto the surface, and observe what happens. Is it the business of mathematics to explain this phenomenon?

14 Non-deductive logic

The logic considered so far has all been *deductive*, which means that if the premises of the reasoning are true, then certainly the conclusion is true. For example, if the premises,

"All men are mortal"

and,

"Socrates is a man"

are true then the conclusion "Socrates is mortal" is certainly also true. Most reasoning in mathematics is of this kind; most reasoning in ordinary life is not. More often, "probability is the guide of life": there is always an element of risk in moving from premises to conclusion. Most real-life inferences are more like,

"90 per cent of courses of drug A improve the patient.
This patient is to be given a course of drug A."

Therefore,

"This patient will improve."

The premises do not make the conclusion certain, they only *support* or *confirm* it. The relation of support or confirmation between propositions (between "evidence" and "hypothesis") is the subject matter of non-deductive logic. It is relevant to such matters as the testing of scientific theories (e.g. "The steady-state theory is less supported by present evidence than the big-bang hypothesis"), proof "beyond reasonable doubt" in law, and statistical inference, such as inferring things about a population from a sample (called inductive inference). Non-deductive logic is not as important in mathematics as it is elsewhere, but it is used sometimes—a mathematician may test a general conjecture by seeing if it is true for particular examples, just as a scientist does with experiments.

Example 1

Find a formula for the sum of the first n odd numbers:

$$1 = 1$$
$$1 + 3 = 4$$
$$1 + 3 + 5 = 9$$
$$1 + 3 + 5 + 7 = 16$$

The answers, 1, 4, 9, 16, are the squares. So we might conjecture that the sum of the first n odd numbers is n^2. This agrees with the data so far. Does further evidence confirm the conjecture?

$$1 + 3 + 5 + 7 + 9 = 25 = 5^2$$
$$1 + 3 + 5 + 7 + 9 + 11 = 36 = 6^2$$

The conjecture is correct in these cases as well, so it is quite well confirmed. It is now clearly worth spending time trying to find a *proof* (we could use mathematical induction, or a technique for adding an arithmetic progression).

Note that checking a hypothesis in a few examples does not *prove* the hypothesis, but it does support it. If a hypothesis is confirmed, the mathematician will be confident enough to set aside valuable time to look for a proof. Confirming an "all" statement by checking some cases of it is called "induction", which must be distinguished from the (deductive) technique of mathematical induction studied in Chapter 8.

Example 2

Euler, the great eighteenth century mathematician, sometimes published his evidence for conjectures before he could prove them. He used some bold steps, which obviously did not constitute a proof, to conclude that the sum of the series,

$$\sum_{n=1}^{\infty} \frac{1}{n^2} = \frac{1}{1^2} + \frac{1}{2^2} + \frac{1}{3^2} + \frac{1}{4^2} + \ldots + \frac{1}{n^2} + \ldots$$

was (the *prima facie* unlikely value) $\pi^2/6$. To confirm this, he calculated both the sum of the series and $\pi^2/6$ to seven decimal places, and found they were equal. Next he used the same method to conclude that,

$$1 - \frac{1}{3} + \frac{1}{5} - \frac{1}{7} + \ldots + \frac{(-1)^{n-1}}{2n - 1} + \ldots = \frac{\pi}{4}$$

This result had already been proved by another method, so that the result obtained by the original method was confirmed. Any rational person should be convinced at this stage that the result is true, "beyond reasonable doubt", even though there is no proof. Euler did later find a full proof of the result.

There is no known way of assigning a number which will indicate the probability of such a confirmation being correct; for example we have no way of saying that on the evidence Euler had, he could be 99 per cent sure that $\pi^2/6$ was the right answer. Now that computers can check large numbers of cases very quickly, it is possible to check some conjectures quite thoroughly. For this reason non-deductive logic has recently become more important in mathematics. (Further reading: G. Polya, *Mathematics and Plausible Reasoning* (1954), Princeton University Press, Princeton, N.J.)

Non-deductive support has been considered, in particular, for some of the long-standing unproved conjectures of mathematics. Probably the best known of these is Fermat's Last Theorem. This states: It is impossible to find positive integers x, y, z and n, with $n \geqslant 3$, such that,

$$x^n + y^n = z^n$$

with x, y, z all non-zero (although there are many solutions with $n = 2$, such as $3^2 + 4^2 = 5^2$). Pierre Fermat asserted this in about 1637, and wrote in the margin of his copy of Diophantus' *Arithmetic*, "I have discovered a truly marvellous demonstration which this margin is too narrow to contain". He never produced any such proof, and no-one has found one since. Fermat did, however, prove the result for $n = 4$. It was found that if there is a counterexample to Fermat's Last Theorem, this implies certain conditions on n which can be checked by calculation. Using this result, it has been checked by computer that Fermat's Last Theorem is true for all values of n up to 125,000. The mathematical community is fairly firmly convinced that the result is true. Anyone who proves (or disproves) it will achieve instant fame.

Exercises

(Gradings of exercises: * easy, ** moderate, *** difficult.)

***1.** Calculate $\frac{1}{1.2} + \frac{1}{2.3} + \frac{1}{3.4} + \ldots + \frac{1}{n(n+1)}$ for the first few values of n. Make a conjecture. Check it for the next value of n. Prove it.

****2.** Calculate $1^2 + 2^2 + \ldots + n^2$ for the first few values of n. Conjecture a formula for $1^2 + 2^2 + \ldots + n^2$. Check if this formula is correct for the next value of n. If so, prove the formula by mathematical induction.

****3.** Do the same for $1^3 + 2^3 + \ldots + n^3$.

****4.** Which positive integers can be expressed in the form $3m + 4n$, where m and n are non-negative integers? Gather some evidence, make a conjecture and prove it.

****5.** Draw a line segment AB and divide it into smaller segments by dots that you label either A or B. Count the number of segments AA, AB, BA and BB. Make some conjecture and prove it.

15

Axioms, symbolic logic and insight

Axioms in geometry

In mathematics the aim is to prove statements from other statements which have already been proved, or from statements which are obvious. Ultimately all statements should be proved from statements which are obvious. It is clear that not everything can be proved: there must be some statements so basic that they cannot be proved from anything simpler. Aristotle said, "It is impossible that there should be demonstration of absolutely everything (there would be an infinite regress, so that there would be no demonstration)".

Besides writing the first book on logic, Aristotle set out in his *Posterior Analytics* (in about 350 B.C.), a program that has in many ways survived to the present day. He believed that each science, but especially mathematics, should begin with "axioms"—a small number of statements which were obviously and necessarily true. Logic should then be used to deduce all the truths of the science from the axioms. This would ensure that all the results deduced were in fact true, and would also show *why* they were true, since it could be seen how they followed from the axioms.

In Aristotle's time, geometry had almost reached this ideal, and appeared to have achieved it not long after, in Euclid's *Elements* (written about 300 B.C.). This has always been recognised as one of the classic works of civilization. It was studied by medieval university students, by schoolboys in the Victorian era, before they set out to govern India, and, was studied, in a modified version, in New South Wales high schools until about 1970. It is now being partially reintroduced. Euclid began by stating certain "definitions", "common notions" and "axioms". There were 23 definitions, some of which were:

1. A *point* is that which has no part.
2. A *line* is length without breadth.
15. A *circle* is a plane figure contained by one line, such that all straight lines falling upon it from one point among those lying within the figure are equal to one another.
16. And the point is called the centre of the circle.

Next there were five postulates or axioms:

1. To draw a straight line from any point to any point.
2. To produce a finite straight line continuously in a straight line.

3. To describe a circle with any centre and any distance.
4. That all right angles are equal to one another.
5. That, if a straight line falling on two straight lines makes the interior angles on the same side less than two right angles, the two straight lines, if produced indefinitely, meet on that side on which are the angles less than the two right angles.

The Axioms 1–3 above are not strictly statements at all, they are instructions instead. However, it is clear that Euclid meant them to be taken as "there exists" statements. Thus Axiom 1 says, "For any two points, there exists a straight line from one to the other."

Then there were five common notions:

1. Things which are equal to the same thing are also equal to one another.
2. If equals are added to equals the wholes are equal.
3. If equals are subtracted from equals, the remainders are equal.
4. Things which coincide with one another are equal to one another.
5. The whole is greater than the part.

Euclid went on to derive a large number of propositions of geometry from these definitions, axioms and notions. The first two are:

Book I, Proposition 1: On a given finite straight line to construct an equilateral triangle.
Book I, Proposition 2: To place at a given point (as an extremity) a straight line equal to a given line.

The Pythagorean theorem is Proposition 47 of Book I. There are many truths of number theory also included. Euclid's proof that there are infinitely many prime numbers appears in Book IX, Proposition 20. Anyone interested in logic should read at least the first few propositions of Book I and follow their proofs. Hundreds of editions of Euclid's *Elements* have been printed. (Thomas L. Heath (ed.), *The Thirteen Books of Euclid's Elements* (1956), Dover, New York, is recommended.)

Since Euclid's time, there have been certain changes in the way axioms are regarded. As Aristotle had hoped, many fields other than geometry came to be organised as systems of truths derived logically from a small number of axioms (e.g. Newtonian mechanics, special relativity, number theory and logic itself). On the other hand, outside of mathematics it proved impossible to find axioms that were self-evident and necessarily true. For example, to find whether the Newtonian axiom, that a body continues to move with constant velocity unless a force acts on it, is true, it is necessary to "get out in the wet" and observe bodies. If the observations agree with the law, the law is confirmed (but not proved). Thus, scientific laws are capable of non-deductive confirmation, but not of proof.

The debate over axioms

In mathematics itself, the status of axioms has become a matter of some controversy. The discovery of non-Euclidean geometries in about 1830 created a sensation. It was realised that from a purely logical point of view, systems of geometry other than Euclid's were just as good. (More precisely, it was proved that non-Euclidean geometries are consistent if and only if Euclidean geometry is consistent.) It follows that it is for measurement to decide whether the space we live in satisfies the Euclidean axioms or not. Ordinary experience shows that our space is approximately three-dimensional Euclidean (at least in our region) but we recognise that it is our *experience* that shows us this, and it is not a necessary and self-evident truth.

Following the discovery of non-Euclidean geometry, an alternative view of axiomatisation became popular. According to this view, axioms are not true or false in mathematics. Rather, the mathematician's job is to find out what consequences follow from what axioms, and it is for someone else, the scientist, to see whether any of these axiom systems fit the real world. Indeed, proponents of this view of mathematics usually say that axioms do not, strictly speaking, even have meaning. In the words of the famous mathematician, David Hilbert:

> "One must be able to say at all times—instead of points, straight lines, planes—tables, chairs and beer mugs."[1]

Mathematics, according to this view, does not study anything at all, but is only the technique of deducing consequences from axioms. As Bertrand Russell expressed it in one of the most widely quoted sayings about mathematics,

> "Mathematics can be defined as the subject in which we never know what we are talking about, nor whether what we are saying is true."[2]

This opinion fitted in well with the scepticism, irrationalism and general unwillingness to be committed, which became popular this century. On the whole, however, it did not make much impression on the practice of mathematicians, who continued to believe that, "$2 + 2 = 4$" is a necessarily true statement about real things, and that the definition of continuity expressed something true about real things that are continuous. Rival geometries were accepted; rival number systems were not. Peano's axioms for the natural number system (1889) were usually regarded as obviously true facts about numbers. His axioms are:

1. 0 is a natural number.
2. If x is a natural number, then there is another natural number x' (called the *successor* of x).
3. $0 \neq x'$ for any natural number.
4. If $x' = y'$, then $x = y$.
5. If p is a property such that, (a) 0 is p and (b) whenever x is p then x' is p, then all natural numbers are p.

This last axiom, the axiom of mathematical induction, is not completely clear, unless it is specified exactly what is meant by "property". It is possible to specify a language of symbols and say that a "property" is anything expressible in that language. In that case all the ordinary facts of number theory can be shown to follow from the axioms (the commutative law of multiplication, the infinity of the prime numbers, and so on). It follows from the famous work of Gödel (1931), however, that there are truths of number theory that do not follow from the axioms (and that adding more axioms will not make the situation any different). In some sense, the number system is too complex to be captured by axioms and the rules of deduction.

1. C. Reid, *Hilbert* (1970), Spring-Verlag, New York, p. 57.
2. B. Russell, *Mysticism and Logic, and Other Essays* (1950), Allen & Unwin, London, p. 75.

Students should acknowledge that there is no consensus about the status of axioms, about whether mathematical truths are necessarily true, or about what Gödel's results really mean. Students may adopt one of the views on the subject, or work out their own, after studying the issues. Or they may, of course, adopt the most common, but least rational course, of remaining sceptical or uninterested in the whole question.

Symbolic logic

Aristotle and Euclid, it will be recalled, proposed to derive truths from "obvious" axioms using "obvious" logical steps. While the "obviousness" of axioms has become more obscure and debatable, logic has become much better understood. It was already known in ancient times, and perhaps even to Euclid himself, that some of the steps in his reasoning were, strictly speaking, not justified by the axioms, but relied on what was "obvious" from a diagram. For example, in Book I Proposition 1, Euclid drew two circles with centres at A and B with radius AB, as shown in Figure 15.1, and then considered a point of intersection C. The fact that the circles do intersect is indeed obvious from the diagram, but it does not follow from any of the axioms.

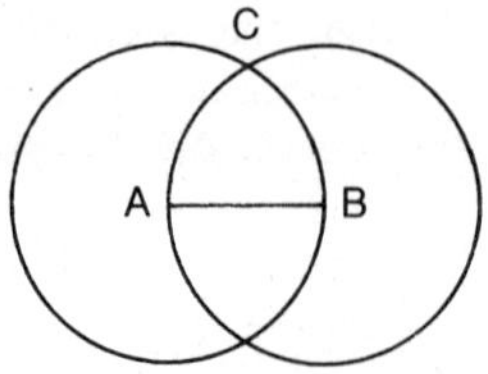

Figure 15.1

During the eighteenth and nineteenth centuries, standards of rigour in proofs increased. In particular, geometrical intuition was found to be occasionally deceptive in the very subtle areas of topology, the foundations of calculus and the theory of continuous functions. Nineteenth-century mathematicians, especially Cauchy and Weierstrass, replaced arguments using geometric intuition with arguments using only logic and very simple concepts about numbers (e.g. the geometric idea of continuity was replaced with the definition given in Chapter 10). The project proved completely successful in avoiding errors in the field, and standards of rigour in proof in these fields have not changed in the last hundred years (though standards in other mathematical fields have tended to catch up to those in the foundations of calculus).

The proofs of Weierstrass had such small logical steps, in the interests of clarity and sureness, that it suggested that the end of the road had been reached in breaking down proofs into smaller and smaller logical steps. The study of this matter by Frege and by Russell and Whitehead led to the creation of the subject of *formal* or *symbolic* logic. They found that all mathematical arguments could be expressed using a very small number of symbols and only two rules of inference:

"Modus ponens": from, "if p then q" and "p", infer "q" (where p and q are any propositions)

and,

"Universal instantiation": from, "for all x, Fx", infer "Fa" (where F is a property and a is the name of an individual).

These steps can be checked mechanically by a computer, so that, in theory, mathematical argument can be expanded into enough steps so that it can be checked by a computer (i.e. without any intuition, though of course intuition is needed to check the validity of the basic steps). In practice, it would take too long to do this for many mathematical proofs (though most of the ones given in this book would only run to a page or so when fully expanded). There are many books on symbolic logic, so no more will be said here. (I. Copi's *Symbolic Logic*[3] is recommended.)

Some work has been done by computer experts on "computer generated proofs", to see if computers could not only check proofs, but also find them. This is in principle something like the problem of teaching computers to play chess. There have been a few successes and computers are just about at the stage of finding and proving the first results considered "interesting". Computers have also been used in proofs that need the checking of a large number of cases. The most famous of these was the four-colour theorem, which stated that any map on a plane can be coloured with at most four colours without any two regions with a common border having thc same colour. The proof of this turned out to need some 1,800 configurations checked and ruled out. The proof given by Appel and Haken in 1977 included the output of 1,200 hours of computer time, and was then checked by another computer.[4]

Aristotle's hope that every result would be proved in such a way as to make the reason for it clear, seems impossible to realise in cases like this.

Insight

This book has concentrated on the mechanical aspects of proof. Often, a proof can be completely constructed using these ideas—the technique of proving an "all" or "some" statement is applied, the definitions of the terms are expanded in symbols, and the proof is complete. Sometimes there is a trick.

Example 1

Prove that $a^n - b^n$ is divisible by $a - b$.

Proof

$$\begin{aligned} a^n - b^n &= a^n - a^{n-1}b + a^{n-1}b - b^n \\ &= a^{n-1}(a - b) + b(a^{n-1} - b^{n-1}) \end{aligned}$$

The proof can then be completed by induction: Assuming the result true for $n - 1$ implies it is true for n, and the result is obviously true for $n = 1$. The first line in this proof, where $a^{n-1}b$ is subtracted from and added to $a^n - b^n$, is simply a trick.

On the other hand, in more interesting cases the idea needed to complete the proof is an insight, or flash of understanding, which reveals some abstract truth hidden in the problem. Below are two examples where the proof consists almost entirely of an insight. Try them before checking the answers.

---

3. I. Copi, *Symbolic Logic*; (1973), (4th edn), Macmillan, New York.
4. K. Appel and W. Haken, "The solution of the Four-Colour-Map Problem", *Scientific American*, (1977), **237**(4), October 1977, pp. 108–21.

Example 2

A chocolate bar is three squares wide and seven squares long. It is broken up into 21 single squares by a sequence of breaks in the usual way—a piece is picked up and broken into two along one of the divisions between squares. Show that no matter in what order the pieces are broken, it always takes the same number of breaks to produce the 21 squares.

Example 3

An ordinary chessboard is eight squares by eight squares, alternately black and white. Two diagonally opposite corner squares are removed. Show that it is not possible to lay down dominoes, with each covering exactly two squares, so that the 62 remaining squares are exactly covered.

Insight can be encouraged or praised, but there is no known way of teaching it, learning it, or programming it into a computer.

Answers

Example 2: Each break divides one piece into two. So after each break there is one more piece than before the break. There is one piece to begin with, so it takes 20 breaks to produce 21 pieces.

Example 3: The two opposite squares are the same colour, so the 62 squares left are not half white and half black. As each domino must cover one white and one black square, the dominoes cannot cover the 62 squares.

Exercises

(Gradings of exercises: * easy, ** moderate, *** difficult.)

**1. "Of a thousand young men who learn logic, no more than ten know anything of the subject six months after the course is finished. That logic is so poorly retained, stems not only from the abstractness and unfamiliarity of the precepts treated in logic, but also from the harnessing of these abstract principles to examples which are unexciting and never encountered elsewhere." (*Logique de Port-Royal*, 1664.)[5]

Discuss.

**2. "I recollect an acquaintance saying to me that, 'the Oriel Common Room stank of Logic'." (Cardinal Newman.)[6]

"Fantasy, my companion, was leaving me south of the border as I crossed into adult life. From now on, I was to face nothing but Reality—a jungle

---

5. A. Arnauld, *The Art of Thinking* (1964), (trans. J. Dickoff and P. James), Bobbs-Merrill, Indianapolis, Ind., p. 21.
6. J. H. Newman, *Apologia Pro Vita Sua* (1964), Oxford University Press, London, p. 176.

leading to the desert plains of Logic." (Anna-Maria dell'Oso.)[7]

Compare and contrast the two statements.

****3.** "Nothing pleased the Chinese as much as the volume on the Elements of Euclid. This perhaps was due to the fact that no people esteem mathematics as highly as the Chinese, despite their method of teaching, in which they propose all kinds of propositions but without demonstrations. The result of such a system is that anyone is free to exercise his wildest imagination relative to mathematics without offering a definitive proof of anything". (Journal of Matteo Ricci, first Jesuit missionary in China)[8]

Discuss.

****4.** "Mathematics is of two kinds, Rigorous and Physical. The former is Narrow: the latter Bold and Broad. To have to stop to formulate rigorous demonstrations would put a stop to most physico-mathematical enquiries. Am I to refuse to eat because I do not fully understand the mechanism of digestion?" (Oliver Heaviside)[9]

Discuss.

---

7. Anna-Maria dell'Oso in *Sydney Morning Herald Good Weekend*, 13 September 1986, p. 38.
8. *China in the Sixteenth Century: The Journals of Matthew Ricci* (1953), (trans. L. J. Gallagher), Random House, New York, p. 476.
9. D. A. Edge, "Oliver Heaviside—Physical Mathematician", *Teaching Mathematics and its Applications*, (1983), **2**, pp. 55–61.

Appendix

Application to computer programs

As computers became commercially available in the late 1960s, it was realised that writing correct computer programs was much harder than had initially been thought. There were huge cost overruns; computers continually sent people bills for $999999.99; and, "there's a problem with the computer", became an excuse for almost anything. It was the time of the "software crisis". There have been three main lines of attack on this problem. One has been to design programs in ways which make them easier to understand, and hence easier in which to find problems. A second line of attack has been to study methods of designing test cases to run on programs that will, with a high probability, reveal the errors in a program. These two methods have improved software reliability enormously, but they do not guarantee that programs are error-free. Since there probably exist computer programs capable of beginning a nuclear war, the highest standards are needed. A third idea has been to *prove* that computer programs do what they are intended to do. Such methods have not yet been applied very well to large programs, but research is continuing and propects seem good. Of course the result will still not be an *absolute* guarantee of correctness, since even proofs can contain mistakes—"to err is human". In this Appendix we provide an outline of one of the techniques used.

A computer works by using a *program* (a sequence of instructions) to assign numbers to various storage locations. These locations are named by letters or words; for example, in a program "a" is used to mean "the number currently in the storage location a". Thus "a > b" means, "the number in a is greater than the number in b". The most basic instruction in programs is the *Assignment Statement*:

$$r \leftarrow 3$$

which means "put 3 in r" or "assign 3 to r".

The instruction, "erase the number in r and replace it with the number in a" (so that number is now in two places, r and a) is denoted:

$$r \leftarrow a$$

"Replace the number in s with the sum of the numbers in s and i" is denoted:

$$s \leftarrow s + i$$

Often, a computer program accomplishes its task by performing assignments of a certain kind *while* a certain condition remains true. For example, to add up the numbers 1 to n, a program should keep adding successive numbers while the numbers are less than or equal to n. A program to do this might consist of the following sequence of instructions (which would be written in some programming language such as BASIC, FORTRAN or PASCAL):

```
Accept a positive integer n typed in by the user
sum ← 1
i ← 1
WHILE i < n DO i ← i + 1, sum ← sum + i
STOP
```

The "WHILE . . . DO" instruction tells the computer to keep increasing i by 1 and adding i to sum as long as i remains less than n. Since i is initially 1 and keeps increasing, it will eventually reach n. The condition $i < n$ is then false and the computer moves on to the next instruction. The, "WHILE . . . DO" instruction is called a "loop", since the computer repeatedly "goes around the loop" of checking whether i is still less than n and adding 1 to i and i to sum.

A program, like a proof, can achieve generality by dealing with all cases at once—the single program shown above deals with *all* positive integers n. We can now ask for a *proof* that the program does what it is intended to do, that is, a proof of the statement:

"For all positive integers n, the above program stops with $1 + 2 + 3 + \ldots + n$ in sum"

This is an "all" statement, so we use the usual method of proof described in Chapter 2.

Proof

Let n be any positive integer. Consider the equation,

$$\text{sum} = 1 + 2 + 3 + \ldots + i$$

This equation is true after the initial assignments,

```
sum ← 1
  i ← 1
```

since then both sides of the equation are 1.

We next show that this equation remains true all the way through the program operation. That is, we show that each time around the loop, although i and sum both change, the relation between them, expressed by the equation,

$$\text{sum} = 1 + 2 + \ldots + i$$

remains true.

Adding $i + 1$ to both sides of this equation gives,

$$\text{sum} + i + 1 = 1 + 2 + \ldots + i + (i + 1)$$

But the left-hand side is the new sum, and the right-hand side is,

$$1 + 2 + \ldots + (\text{new})i$$

since each time round the loop 1 is added to i and $i + 1$ (the new i) is added to sum.

So the equation,

$$\text{sum} = 1 + 2 + \ldots + i$$

is still true for the new values of i and sum.

Therefore, the equation,

$$\text{sum} = 1 + 2 + \ldots + i$$

is true before the "WHILE . . . DO" instruction begins acting, and remains true as each loop is performed. Since i is initially 1 and increases by 1 during each loop, it is eventually equal to n. At this stage the condition $i < n$ is false for the first time, so the computer stops with,

$$\text{sum} = 1 + 2 + \ldots + n$$

as required.

An equation such as sum = 1 + 2 + . . . + i, which remains true as a program goes around a loop, is called "loop invariant". For further reading on this topic, *The Science of Programming* by D. Gries[1] is recommended.

Exercises

(Gradings of exercises: * easy, ** moderate, *** difficult.)

*1. Write a program similar to the one in the text (p. 151) which finds, for any positive integer n, the sum of the first n even numbers. Prove that it does perform as required.

**2. To make computers perform complex operations, it is necessary to define the complex operations in terms of simpler ones. One way of defining division of positive integers is to use repeated subtraction of the smaller from the larger: the number of subtractions needed to get to zero is the quotient. The following program is designed to do this:

```
Accept two integers A and B, with A > B > 0
a ← A
b ← B
q ← 0
WHILE a > 0 DO q ← q + 1, a ← a − b
IF a < 0 PRINT "A is not divisible by B", STOP
IF a = 0 PRINT "The quotient of A and B is", PRINT q, STOP
```

(a) Show that the equation,

$$A = a + qb$$

is a loop invariant, and hence show that if $A = QB$ for some positive integer Q, then the program stops with the quotient Q printed out.

---

1. D. Gries, *The Science of Programming* (1981), Springer-Verlag, New York.

(b) Show that if A is not divisible by B, then the message "A is not divisible by B" will be printed.

(c) Suppose that in the first line we replace "A > B > 0" by "A > B" and that we then type in a positive integer for A and 0 for B. Show that the program never stops.

****3.** A program has the following instructions:

```
Accept an integer N typed in by the user
n ← N
WHILE n is even DO n ← n/2
IF n is odd PRINT n
STOP
```

Show that for most integers N, the program prints out an odd number. Are there any exceptions?

****4.** The proofs in Questions 1, 2 and 3 above and the proof in the text claim to prove something about *all* positive integers. But computers cannot store arbitrarily large integers. Discuss the significance of this problem for the proofs.

Solutions to selected exercises

Chapter 1

1.
$$\frac{1}{1000} - \frac{1}{1002} = \frac{1002}{1000 \times 1002} - \frac{1000}{1000 \times 1002}$$
$$= \frac{2}{1000 \times 1002}$$
$$= \frac{2}{1002000}$$

But, 1002000 > 1000000, so,

$$\frac{2}{1002000} < \frac{2}{1000000}$$

Therefore, $\frac{1}{1000} - \frac{2}{1002} < \frac{2}{1000000}$

5.
$$\tan\left(\frac{\pi}{12}\right) = \tan\left(\frac{\pi}{3} - \frac{\pi}{4}\right)$$
$$= \frac{\tan\left(\frac{\pi}{3}\right) - \tan\left(\frac{\pi}{4}\right)}{1 + \tan\left(\frac{\pi}{3}\right)\tan\left(\frac{\pi}{4}\right)}$$
$$= \frac{\sqrt{3} - 1}{1 + \sqrt{3}(1)}$$
$$= \frac{(\sqrt{3} - 1)(\sqrt{3} - 1)}{(\sqrt{3} + 1)(\sqrt{3} - 1)}$$
$$= \frac{3 - 2\sqrt{3} + 1}{3 - 1}$$
$$= \frac{4 - 2\sqrt{3}}{2} = 2 - \sqrt{3}.$$

6. Putting $x = 1 - \sqrt{5}$ in the left hand side of the equation,

$$(1 - \sqrt{5})^3 - 3(1 - \sqrt{5})^2 - 2(1 - \sqrt{5}) + 4$$
$$= 1 - 3\sqrt{5} + 3(5) - 5\sqrt{5} - 3 + 6\sqrt{5} - 15 - 2 + 2\sqrt{5} + 4 = 0.$$

10. The accuracy of the answer is ridiculous; the diameter is given only to one significant figure, and ponds are not perfectly circular.

12. On squaring both sides we would have,

$$(2 + \sqrt{2}) + 2\sqrt{2 + \sqrt{2}}\sqrt{2 - \sqrt{2}} + (2 - \sqrt{2}) < 4(2)$$

Rearranging, we obtain,

$$2\sqrt{(2 + \sqrt{2})(2 - \sqrt{2})} < 4$$

giving,

$$\sqrt{2} < 2$$

which is true. Each step is reversible, so the final proof consists of the above steps in the reverse order.

17. The first few steps are: Dividing through by $\sqrt[3]{3}$,

$$\sqrt[3]{1 + 3^{-2/3}} + \sqrt[3]{1 - 3^{-2/3}} < 2$$

Then subtract $\sqrt[3]{1 + 3^{-2/3}}$ from both sides and consider the cube of the right-hand side.

20.
$$\frac{100^{99}}{99^{100}} = \left(\frac{100}{99}\right)^{99} \frac{1}{99}$$
$$= \left(1 + \frac{1}{99}\right)^{99} \frac{1}{99}$$
$$< 3.\frac{1}{99}$$

(since, $\left(1 + \frac{1}{99}\right)^{99} < 3$, see problem 18) and this is less than 1.

So,

$$99^{100} > 100^{99}$$

Chapter 2

2. Let n and $n + 1$ be any pair of consecutive integers. Then,

$$n + (n + 1) = 2n + 1$$

which is odd.

5. True.

$$x^2 - 3x + 2 = (x - 1)(x - 2)$$

This is negative for all x between 1 and 2, since the factor $(x - 2)$ is negative and the factor $(x - 1)$ is positive, so their product is negative.

6. (b) No. (But fathers who use misleading language may find they have ungrateful children.)

10. The equation,

$$\frac{n}{n + 1} < \frac{n + 1}{n + 2}$$

is equivalent to,

$$n(n + 2) < (n + 1)^2$$

(since n + 1 and n + 2 are positive).

That is,

$$n^2 + 2n < n^2 + 2n + 1$$

This is true, so the final proof consists of these steps in the reverse order.

12. Let the middle number be 2n + 1.

Then the product of the three consecutive numbers is 2n(2n + 1)(2n + 2). One of these three is divisible by 3; of the two consecutive even numbers 2n and 2n + 2, one is divisible by 4 and the other by 2.

So the product is divisible by 3 × 4 × 2 = 24.

15. Since x is positive, we can multiply both sides of the inequality x < 1 by x. This gives the answer.

18. (a) This can be done by substituting the two numbers given for x in $ax^2 + bx + c$ and showing that the result is 0.
(b) This can be done by completing the square.

19. One generalisation is,

$$\frac{1}{n} - \frac{1}{n + 2} < \frac{2}{n^2}$$

20. Since 7 divides a + b, it divides,

$$(a + b)^2 = a^2 + 2ab + b^2$$

Hence 7 also divides the difference of this and $a^2 + b^2$, which is 2ab.

But 7 does not divide 2, so it must divide ab.

Since 7 is a prime, it must divide either a or b.

But if it divides a, it divides b as well (since it divides a + b); similarly, if 7 divides b, it also divides a.

So it divides both a and b.

22. Both the cosine rule and the equation,

$$\cos^2\theta + \sin^2\theta = 1$$

rely on Pythagoras' theorem for their proof.

So they cannot be used to prove Pythagoras' theorem, since this would make the argument circular.

26. A proof is given in Euclid's *Elements*, Book 1, proposition 47.

29. Adding the equations gives,

$$2x = A + B$$

so,

$$x = \tfrac{1}{2}(A + B)$$

Substituting this in the first equation gives,

$$y = A - x = \tfrac{1}{2}(A - B)$$

So the equations can be solved, no matter what A and B are.

33. Equating the first, second and third co-ordinates on both sides, we have the system of equations,

$$\begin{aligned} x + y + 3z &= 0 \\ x + 2y + 4z &= 0 \\ 3y + 3z &= 0 \end{aligned}$$

Solving these by row reduction, we find $x = -2\lambda$, $y = -\lambda$, $z = \lambda$.

So,

$$(x, y, z) = -\lambda(2, 1, -1)$$

which is a multiple of $(2, 1, -1)$.

34. The graphs of $y = x$ and $y = \sin x$ meet at $x = 0$. The gradient of $y = x$ is 1 (always), while the gradient of $y = \sin x$ is $\cos x$.
But, $\cos x \leqslant 1$, and is sometimes less than 1. So the graph of $y = \sin x$ lies below $y = x$ for all $x > 0$.

38. True. The problem cannot be done straightforwardly by integrating $(\sin x)/x$, since this function does not have an integral expressible in terms of elementary functions. It can be done by comparing areas; draw a rough graph first.

Chapter 3

1. Yes. (In fact we could *prove* that they are equivalent, by showing that each implies the other.)

4. (a) True.
   (b) False, since 5 is an odd prime.

5. (c).

6. The proof is in two parts:

   First part: To show that if x is odd then x^2 is odd.
   If x is odd, then $x = 2k + 1$ for some integer k.
   So $x^2 = 4k^2 + 4k + 1$, which is odd.

   Second part: To show that if x^2 is odd then x is odd.
   Suppose x^2 is odd.
   If x were even, then x^2 would be even, contrary to what we have just supposed.
   So x cannot be even, so it must be odd.

9. Let x be a number with digits $a_n, a_{n-1}, \ldots, a_1, a_0$. Then,

$$x = 10^n a_n + \ldots + 100a_2 + 10a_1 + a_0$$

The number formed by the last two digits is $10a_1 + a_0$, and $x - (10a_1 + a_0)$ is a multiple of 100, which is a multiple of 4.

Hence if x is divisible by 4, so is $10a_1 + a_0$, and if $10a_1 + a_0$ is divisible by 4, so is x.

14. Suppose that $AD - BC = 0$.

Case 1: Neither B nor D is zero.

Then since $AD = BC$, we obtain,

$$A/B = C/D$$

and hence,

$$-A/B = -C/D.$$

Thus, the gradients of the two lines are equal, so they are parallel.

Case 2: Either B or D is zero.

If $B = 0$, then $AD = 0$, so either $A = 0$ or $D = 0$.

If $A = 0$ then the first "line",

$$Ax + By + E = 0$$

would be $E = 0$. There is no such line so we must have $D = 0$.

Then the two lines are,

$$Ax + E = 0$$

and,

$$Cx + F = 0$$

which are both vertical, and hence parallel.

A similar argument applies if $D = 0$. Hence in both cases, the two lines are parallel.

Conversely, suppose that the lines,

$$Ax + By + E = 0$$

and,

$$Cx + Dy + F = 0$$

are parallel.

Case 1: Neither B nor D is zero.

Then the gradient of the first line is $-A/B$ and the gradient of the second line is $-C/D$.

These are equal since the lines are parallel, so,

$$-A/B = -C/D$$

and hence,

$$AD - BC = 0$$

Case 2: Either $B = 0$ or $D = 0$.

If $B = 0$ then the first line is,

$$Ax + E = 0$$

which is vertical.

So the other line is vertical too, that is, $D = 0$.

Thus AD − BC = 0 − 0 = 0.

Similarly, if D = 0 then B = 0 and again AD − BC = 0.

18. Suppose that x has a decimal expansion that is terminating or (eventually) repeating. If the expansion is terminating,

$$x = a_n a_{n-1} \ldots a_1 . b_1 \ldots b_m$$

So,

$$x = \frac{a_n \ldots a_1 b_1 \ldots b_m}{10^m}$$

which is rational. If x has an (eventually) repeating decimal,

$$x = a_n \ldots a_1 . b_1 \ldots b_{r-1} \overline{b_r \ldots b_m}$$

Then x = y + z, where,

$$y = a_n \ldots a_1 . b_1 \ldots b_{r-1}$$

(which is rational), and,

$$z = .00\ldots00\overline{b_r \ldots b_m}$$

Then,

$$10^{r-1} z = 0.\overline{b_r \ldots b_m}$$

and,

$$10^{m-r} 10^{r-1} z = b_r \ldots b_m . \overline{b_r \ldots b_m}$$

Subtracting the last two equations,

$$10^{m-1} z - 10^{r-1} z = b_r \ldots b_m$$

giving,

$$z = b_r \ldots b_m / (10^{m-1} - 10^{r-1})$$

which is rational. Thus, x is rational.

Conversely, suppose x is rational, so that x = m/n for some integers m and n.

Let us obtain the decimal expansion by dividing n into m.

Let the remainder produced by the division step that gives the kth decimal place be p_k.

p_k is an integer less than n.

After at most n division steps some p_k will be repeated. Once p_k reappears then the string of digits between the appearances of p_k will be repeated.

Chapter 4

1. (a) By finding an integer x such that 7073 = 643x.
   (b) 7073 = 643 × 11.

2. Yes, 31 and 42.

4. 6 is a perfect number.

6. (a) True: if x is positive, x^2 is positive; if x is negative, x^2 is also positive; if x = 0 then $x^2 = 0$. So in all cases, $x^2 \geqslant 0$.

(b) $1^2 > 0$, so for some $x \in \mathbf{R}$, $x^2 > 0$.

7. (a) T (b) F (c) T (d) T (e) F (f)T (g) T (h) F.

8. $2^{1/4}$ is such a number.

9. (a) F (b) F (c) T (d) F.

10. All are true.

13. N = 20 (or any larger integer) satisfies the condition.

14. The line y = x is an axis of symmetry, since if (x, y) lies on the curve, then,

$$x^2 + xy + y^2 = 1$$

Thus,

$$y^2 + yx + x^2 = 1$$

so the point (y, x) also lies on the curve; this is the point opposite (x, y) across the line y = x.

17. (a) $(1, 0, 1) = 1(0, 0, 1) + (-1)(0, 1, 1) + 1(1, 1, 1)$
    (b) No, because any linear combination of the four given vectors has last coordinate zero, but (1, 2, 1) does not.
    (c) $12(3, 4) + 1(-36, -48) = (0, 0)$.

18. Yes, for example, (3, 1, 0).

21. Yes, for example, e^{-x}.

Chapter 5

1. (a) Two, both existential.
   (b) True: m = 1, n = −1 is one solution.

2. This is similar to Chapter 2, Exercise 18(a), but the case a = 0 must be considered separately.

3. We want,

$$\frac{1}{N} - \frac{1}{N+1} < \frac{1}{n}$$

that is,

$$\frac{1}{N(N+1)} < \frac{1}{n}$$

Now,

$$\frac{1}{N(N+1)} < \frac{1}{N^2}$$

So if we can find N such that,

$$\frac{1}{N^2} < \frac{1}{n}$$

then,

$$\frac{1}{N(N + 1)}$$

will also be less than $1/n$.

But,

$$\frac{1}{N^2} < \frac{1}{n}$$

is equivalent to $N^2 > n$, that is, $N > \sqrt{n}$.

So any integer N greater than $\sqrt{n}$ will satisfy the condition.

4. (a) Suppose $ad - bc \neq 0$. Then a times the second equation minus c times the first equation gives,

$$(ad - bc)y = af - ce$$

and hence,

$$y = (af - ce)/(ad - bc)$$

(the division is possible since $ad - bc \neq 0$).
Similarly,

$$x = (bf - ed)/(ad - bc)$$

So there exists a solution to the system.

(b) No, it is not true that if there exists a solution then $ad - bc \neq 0$.
For example, the equations $x + y = 1$ and $2x + 2y = 2$.

(c) No; the same example applies (in fact, these two questions are logically equivalent, as will be explained in Chapter 6).

7.

$$n^5 - n = n(n^4 - 1) = n(n - 1)(n + 1)(n^2 + 1)$$

5 divides one of the five consecutive numbers $n - 2, n - 1, n, n + 1, n + 2$.

If it divides $n - 1$, n or $n + 1$, then it divides $n^5 - n$.

If 5 divides $n - 2$, then,

$$n = 5k + 2$$

for some k, so,

$$n^2 + 1 = 25k^2 + 20k + 5$$

which is a multiple of 5; thus 5 again divides $n^5 - n$.

A similar argument holds if 5 divides $n + 2$. (A proof by mathematical induction is also possible.)

8. For any positive real number d there exists an integer n such that $n > 1/d$. So $1/n < d$.

We now take any two irrational numbers a and b, and take $d = b - a$.

So there exists an integer n such that $1/n < b - a$.

We consider the set of rational numbers of the form h/n (h an integer), and show that one of them must lie between a and b.

Let h_1 be the largest integer such that $h_1/n < a$. So,

$$h_1/n < a < (h_1 + 1)/n.$$

Now,

$$(h_1 + 1)/n - a < (h_1 + 1)/n - h_1/n = 1/n < b - a$$

so,

$$a < (h_1 + 1)/n < b$$

Thus, there is a rational number between a and b.

11. Take one of the points, A. It is joined to five others, so at least three of the lines from A are of the same colour, say red. Call the points at their ends B, C and D. If one of BC, BD, CD is red, then there is a red triangle; if not, there is a blue triangle, BCD.

12. We have to find whether, for all (x, y, z), there exist scalars a, b, c such that,

$$(x, y, z) = a(1, 0, 0) + b(1, 1, 0) + c(1, 1, 1)$$

Solving, we find,

$$(x, y, z) = (x - y)(1, 0, 0) + (y - z)(1, 1, 0) + z(1, 1, 1)$$

So there exist such scalars a, b, c.

16. Yes, for example

$$f(x) = e^{-\frac{1}{x^2}}$$

(with f(0) defined to be 0).

Chapter 6

1. Both require one counterexample to be found. For example, "Not all prime numbers are odd", is proved true by finding the counterexample 2, which is an even prime.

   The statement, "All prime numbers are odd", is proved false by exhibiting the same counterexample.

4. (a) True.
   (b) False.

5. (a) "My car is not red".
   (b) (i) "Not all cars are red".
   (ii) "Some cars are not red".
   (c) "For all $x \in \mathbf{R}$, $x^2 \neq 2$".

8. False; take $a = 1$, $b = 2$, $c = 0$. Then $ac = bc$ but $a \neq b$.

10. (a) Take $x = 0$.
    (b) Take 6 itself.

11. (b) False; $-2 < -1$ and $-4 < -3$ but $(-2)(-4) > (-1)(-3)$.

15. If n is even, then $n = 2k$ for some integer k.

    So $n^2 = 4k^2$, which is a multiple of 4.

    If n is odd,

$$n = 2k + 1$$

so,

$$n^2 = 4k^2 + 4k + 1$$

which is 1 plus a multiple of 4.

So for all integers n, n^2 is either a multiple of 4 or 1 more than a multiple of 4, so it cannot be of the form $4k + 2$ or $4k + 3$.

17. First,

$$a + \sqrt{b} - c = \sqrt{d}$$

Squaring and rearranging gives,

$$2(a - c)\sqrt{b} = d - b - (a - c)^2$$

So $a - c$ must be zero, since otherwise the left-hand side would be irrational and the right-hand side rational. Putting this in the original equation results in $\sqrt{b} = \sqrt{d}$, so $b = d$.

19. One of the three consecutive integers $2^n - 1, 2^n, 2^n + 1$ is a multiple of 3. It cannot be 2^n, since 3 does not divide 2.

So 3 divides either $2^n - 1$ or $2^n + 1$.

So these are not primes, unless one of them is 3. This happens when $n = 1$ or 2.

23. Yes. Both are equivalent to, "There is nothing that is both an A and a B".

25. We prove the contrapositive: If,

$$f(x) = f(y)$$

then,

$$mx + b = my + b$$

so $mx = my$,

so $x = y$ (since $m \neq 0$).

Therefore, if $x \neq y$ then $f(x) \neq f(y)$.

26. By calculating enough angles, we find that four of the angles in the octagon are $\pi/2 + 2\tan^{-1}\frac{1}{2}$ while the other four are $\pi - 2\tan^{-1}\frac{1}{2}$. It can be checked that these angles are not equal using a proof by contradiction.

30. No; for example, take $\underset{\sim}{v}_1 = (1, 0)$, $\underset{\sim}{v}_2 = (2, 0)$, $\underset{\sim}{v}_3 = (0, 1)$.

Then $\underset{\sim}{v}_3$ is not a linear combination of $\underset{\sim}{v}_1$ and $\underset{\sim}{v}_2$, but $\underset{\sim}{v}_1$, $\underset{\sim}{v}_2$ and $\underset{\sim}{v}_3$ are dependent.

32. Yes. For example, $a = b = c = d = 1$.

34. No. For example, $f(x) = |x|$.

35. No. For example, $f(x) = e^x$.

36. (b) No. For example, $f(x) = x^4$ has a minimum at $x = 0$, but its second derivative there is zero.

Chapter 7

1. (a) "All members of A are members of B"; "If something is a member of A then it is a member of B".
   (b) Hence the proof of "$A \subset B$" must begin by assuming $x \in A$ and must deduce from this $x \in B$.

3. Suppose that $A \subset B$ and $B \subset C$.

   Let $x \in A$. Therefore $x \in B$ (since all members of A are members of B).

   Hence $x \in C$ (since all members of B are members of C).

   So we have proved that if $x \in A$ then $x \in C$.

   Therefore, $A \subset C$.

9. (b) Yes

12. (b) Let $x \in \overline{(A \cup B)}$.

    So $x \notin A \cup B$.

    So $x \notin A$ and $x \notin B$ (since if x were in either A or B, it would be in $A \cup B$).

    So $x \in \bar{A}$ and $x \in \bar{B}$.

    So $x \in \bar{A} \cap \bar{B}$.

    This shows that, $\overline{(A \cup B)} \subset \bar{A} \cap \bar{B}$.

    The proof of the converse consists of the same steps in the reverse order.

13. The reasoning is correct.

14. (a) The set of real numbers itself is not bounded.
    (b) Suppose that S is bounded and that $T \subset S$.

    Then there exist numbers M and N such that for all $s \in S$, $M < s < N$.

    Now let $t \in T$.

    Since $T \subset S$, $t \in S$, so $M < t < N$.

    So for all $t \in T$, $M < t < N$. That is, M and N are also bounds for T.

    Therefore, T is bounded.
    (c) Let $\{a_1, \ldots, a_n\}$ be a finite set of real numbers. Let a_i be the minimum of these, and a_j be the maximum.

    Then if we take $M = a_i - 1$ and $N = a_j + 1$, then $M < s < N$ for all s in the set.

    So the set is bounded.
    (d) This is logically equivalent to (b), so it is already proved.
    (e) Suppose that S and T are bounded. So there exist M and N such that for all $s \in S$, $M < s < N$.

    Now let $r \in S \cap T$. Then $r \in S$, so $M < r < N$.

    So M and N are bounds for $S \cap T$, so $S \cap T$ is bounded. (Note that the condition "T is bounded" was not used, so we may strengthen the conclusion to: "If S is bounded and T is any set of real numbers, then $S \cap T$ is bounded".)
    (f) Yes. Let S and T be bounded, so that there exist M_1 and N_1 such that $M_1 < s < N_1$ for all $s \in S$, and there exist M_2 and N_2 such that $M_2 < t < N_2$ for all $t \in T$.

Now let M_3 be the minimum of M_1 and M_2, and let N_3 be the maximum of N_1 and N_2.

Take $r \in S \cup T$. Then $r \in S$ or $r \in T$.

If $r \in S$, then $M_3 \leqslant M_1 < r < N_1 \leqslant N_3$.

If $r \in T$, then $M_3 \leqslant M_2 < r < N_2 \leqslant N_3$.

So for all $r \in S \cup T$, $M_3 < r < N_3$.

So $S \cup T$ is bounded.

(g) As in (e), let M_1 and N_1 be bounds for S, and M_2 and N_2 be bounds for T.

Then is it straightforward to show that for all $u \in U$, $M_1 + M_2 < u < N_1 + N_2$.

So U is bounded.

17. $(A \cap B) \cup C = A \cap (B \cup C)$ if and only if $C \subset A$.

20. Take a vector (a, b, c) such that $a + b + c = 0$. Then $c = -a - b$, so the vector is of the form $(a, b, -a - b)$.

So we need to show that any vector $(a, b, -a - b)$ can be written as $x(1, 1, -2) + y(1, 0, -1)$.

Solving the equations gives $x = b$, $y = a - b$.

So the equations can always be solved.

24. All functions of the form $f(x) = x^2 + C$, for $C \in \mathbf{R}$, are solutions, so the set of solutions is infinite.

25. (c) Yes. Any function of the form $C \sinh x + D \cosh x$ is $((C + D)/2)e^x + ((D - C)/2)e^{-x}$, while any function $A e^x + B e^{-x}$ is $(A - B)\sinh x + (A + B)\cosh x$.
So the sets are equal.

Chapter 8

1. (a) For $n = 1$, $1^2 = \frac{1}{6}(1)(2)(3)$, so the formula is true for $n = 1$.
We assume the formula is true for n,

$$1^2 + 2^2 + \ldots + n^2 = \tfrac{1}{6}n(n + 1)(2n + 1)$$

Then,

$$\begin{aligned}
1^2 + 2^2 + \ldots + n^2 + (n + 1)^2 &= \tfrac{1}{6}n(n + 1)(2n + 1) + (n + 1)^2 \\
&= (n + 1)(\tfrac{1}{6}n(2n + 1) + n + 1) \\
&= (n + 1)\frac{2n^2 + n + 6n + 6}{6} \\
&= \frac{(n + 1)(n + 2)(2n + 3)}{6} \\
&= \tfrac{1}{6}(n + 1)((n + 1) + 1)(2(n + 1) + 1)
\end{aligned}$$

This is the formula with n replaced by $n + 1$. So we have shown that the truth of the formula for n implies its truth for $n + 1$.

Since it is true for $n = 1$, it is true for all natural numbers n.

2. For $n = 1$, the expression is 1, which is positive.
Assume that $1 - 1/2 + 1/3 - \ldots + (-1)^{n-1}\, 1/n$ is positive for all n less then or equal to some number k.

If k is even, the expression for $k + 1$ is,

$$1 - \frac{1}{2} + \ldots - \frac{1}{k} + \frac{1}{k+1}$$

which is positive, since it is $1/(k + 1)$ plus something that we have assumed is positive.

If k is odd, the expression for $k + 1$ is,

$$1 - \frac{1}{2} + \ldots - \frac{1}{k-1} + \left(\frac{1}{k} - \frac{1}{k+1}\right)$$

which is positive because $1/k - 1/(k + 1)$ is positive and we have assumed that $1 - 1/2 + \ldots - 1/(k - 1)$ is positive.

So the expression is positive for all n.

(Note that if we bracket the expression as,

$$\left(1 - \frac{1}{2}\right) + \left(\frac{1}{3} - \frac{1}{4}\right) + \ldots + \left(\frac{1}{2m-1} - \frac{1}{2m}\right)$$

It is clear that the expression must always be positive; the above inductive argument simply made the reasoning here explicit.)

7. (b)
$$\frac{1}{2!} + \frac{2}{3!} + \ldots + \frac{n}{(n+1)!} = \frac{(n+1)! - 1}{(n+1)!}$$

8. The reasoning does not work when $n = 1$.

10. (b)
$$s_n = \tfrac{1}{2}n(n + 1) + 1$$
(c) The essential step is to prove that $s_{n+1} = s_n + n + 1$. This is true because the $(n + 1)$st line cuts all the existing n lines, and hence creates $n + 1$ new regions.

14. The result is clearly true for $n = 1$. Assume it is true for a number n. Then,

$$\begin{aligned}\frac{d^{n+1}(x^{n+1})}{dx^{n+1}} &= \frac{d^n}{dx^n}\left(\frac{d}{dx}(x\,x^n)\right)\\ &= \frac{d^n}{dx^n}(x^n + x\,n\,x^{n-1})\\ &= n! + n\left(\frac{d^n}{dx^n}x^n\right)\\ &= n! + n.n! = (n + 1)n! = (n + 1)!\end{aligned}$$

Chapter 9

1. The set does not have a zero (nor is arbitrary multiplication by scalars possible, since, for example, $1/2 \times 1$ is not a positive integer).

2. As an example, we check the commutative law of addition.
Take two vectors in $\mathbf{R}^2$, (x_1, y_1) and (x_2, y_2).

Then,

$$(x_1, y_1) + (x_2, y_2) = (x_1 + x_2, y_1 + y_2)$$

by the definition of addition of vectors.

$$= (x_2 + x_1, y_2 + y_1)$$

by the commutative law of addition of numbers.

$$= (x_2, y_2) + (x_1, y_1)$$

by the definition of addition of vectors.

So we have shown that for all $\underset{\sim}{u}, \underset{\sim}{v} \in \mathbf{R}^2$, $\underset{\sim}{u} + \underset{\sim}{v} = \underset{\sim}{v} + \underset{\sim}{u}$.

3. (a)

$$(-1)\underset{\sim}{u} + \underset{\sim}{u} = (-1)\underset{\sim}{u} + 1\underset{\sim}{u}$$

(since $1\underset{\sim}{u} = \underset{\sim}{u}$).

$$= ((-1) + 1)\underset{\sim}{u}$$

(distributive law).

$$= 0\underset{\sim}{u}$$

(addition of numbers).

$$= \underset{\sim}{0}$$

(using the result proved earlier).

Adding $-\underset{\sim}{u}$ to both sides of the equation gives,

$$((-1)\underset{\sim}{u} + \underset{\sim}{u}) + (-\underset{\sim}{u}) = \underset{\sim}{0} + \underset{\sim}{u}$$
$$(-1)\underset{\sim}{u} + (\underset{\sim}{u} + (-\underset{\sim}{u})) = \underset{\sim}{u}$$

(associative law of addition of vectors, and definition of $\underset{\sim}{0}$).

$$(-1)\underset{\sim}{u} + \underset{\sim}{0} = \underset{\sim}{u}$$

(definition of $-\underset{\sim}{u}$).

$$(-1)\underset{\sim}{u} = \underset{\sim}{u}$$

(definition of $\underset{\sim}{0}$).

6. (1, 0, 0) is in the subset, but 2(1, 0, 0) = (2, 0, 0), which is not in the subset. So the subset is not closed under scalar multiplication, and so is not a subspace of $\mathbf{R}^3$.

8. Let X and Y be matrices in the set. So AX = XA and AY = YA.

   Then A(X + Y) = AX + AY = XA + YA = (X + Y)A. So X + Y also satisfies the condition for being in the set.

   Therefore, the set is closed under addition.

   Now let X be in the set, so that AX = XA, and let λ be a scalar.

   Then $A(\lambda X) = \lambda(AX) = \lambda(XA) = (\lambda X)A$.

   So λX is also in the set.

   Thus, the set is closed under scalar multiplication.

   Therefore, the set is a subspace of M_{22}.

   (Note carefully that the matrix A is fixed throughout the question.)

10. (b) The function f given by f(x) = 2x is in H but 2f is not, since $\int_0^1 2f(x)\,dx = 2$.

Chapter 10

1. (a) 1.
   (b) 10.
   (c) Any X greater than or equal to $1/\sqrt{\epsilon}$.

2. (a) 0.
   (b) 100 or any larger number.
   (c) $1/\epsilon$ or any larger number.
   (d) Just less than π; No.

3. (b) 0.62; about 15 minutes.

4. Take $\epsilon > 0$.

$$\left|\frac{x-1}{2x+1} - \frac{1}{2}\right| = \left|\frac{2(x-1)-(2x+1)}{2(2x+1)}\right| = \left|\frac{-3}{2(2x+1)}\right|$$

This equals, $\dfrac{3}{2(2x+1)}$, which is less than ϵ provided $\frac{3}{2} < (2x+1)\epsilon$, that is,

$$x > \frac{1}{2}\left(\frac{3}{2\epsilon} - 1\right)$$

So,

$$\lim_{x\to\infty} \frac{x-1}{2x+1} = \frac{1}{2}$$

8. (b) $-\frac{1}{3}$.

11. (a) Take $\epsilon > 0$.

$$\left|\frac{1}{2^n} - 0\right| = \frac{1}{2^n}$$

which is less than ϵ provided $2^n > 1/\epsilon$, that is, $n > \log_2(1/\epsilon)$.

So there does exist N such that for all $n > N$,

$$\left|\frac{1}{2^n} - 0\right| < \epsilon$$

We can take N to be the smallest integer greater than $\log_2(1/\epsilon)$.

(b) We use proof by contradiction. Suppose there did exist a limit L. Then it would be true that:

For all $\epsilon > 0$ there exists N such that for all $n > N$, $|2^n - L| < \epsilon$.

We show that this "all" statement must be false by showing that $\epsilon = 1$ is a counterexample. If there did exist N such that for all $n > N$ $|2^n - L| < 1$, then $2^{N+1}, 2^{N+2}, \ldots$ would all be within 1 of the limit L.

This is impossible, since these numbers differ by more than 1.

13. $\text{Min}(\sqrt{4\frac{1}{2}} - 2, 2 - \sqrt{3\frac{1}{2}}) = 0.12$.

19. (c) Redefine $f(2)$ to be 4.

20. (a) $\frac{1}{2}\sin^{-1}0.49 = 15°$ or $\pi/2 - \frac{1}{2}\sin^{-1}0.49 = 75°$.
    (b) We find that for a range R, $\theta = \frac{1}{2}\sin^{-1}(9.8R/100^2)$ (or $\pi/2$ minus this).

    To estimate the change $\Delta\theta$ in θ caused by a change ΔR in R, at $R = 500$, we use $\Delta\theta = \frac{d\theta}{dR}\Delta R$, and find that the accuracy needed in θ is 3.2×10^{-2} degrees, (which seems impossible to attain).

21. Let a be any real number, and $\delta > 0$. Then there are both rational numbers and irrational numbers in the interval $(a - \delta, a + \delta)$.

    So there are points x in the interval $(a - \delta, a + \delta)$ such that $f(x) = 0$, and other points x such that $f(x) = 1$.

    So if we take $\epsilon = \frac{1}{3}$, it is impossible that there should exist δ such that,

$$|f(x) - f(a)| < \epsilon$$

whenever $|x - a| < \delta$. So the statement,

For all $\epsilon > 0$ there exists $\delta > 0$ such that $|f(x) - f(a)| < \epsilon$ whenever $|x - a| < \delta$

is false (since $\epsilon = \frac{1}{3}$ is a counterexample).

Therefore, f is not continuous at a.

23. For x such that $|x| > \sqrt[7]{(7!/10^4)} = 0.9$.

24. (b) Using (a), we find the first n such that $5^n/n! < \frac{1}{2}10^{-4}$, by calculating successively the numbers $5^n/n!$. We find $n = 20$.

Chapter 11

1.

000	$\varnothing$	110	$\{a_1, a_2\}$
100	$\{a_1\}$	101	$\{a_1, a_3\}$
010	$\{a_2\}$	011	$\{a_2, a_3\}$
010	$\{a_3\}$	111	$\{a_1, a_2, a_3\}$.

3. (a) $\binom{14}{3} = 364$.

    (b) Every solution of $a + b + c + d = 7$ with a, b, c, d non-negative integers, gives a solution,

$$(a + 1) + (b + 1) + (c + 1) + (d + 1) = 11$$

and vice versa, so the answer is,

$$\binom{10}{3} = 120$$

    (c) 56.

4. 23; $n - 1$, since there are $n - 1$ losers and one final winner.

5. Hint: we need to find the first number of the form $(n + 1)/2n$ or $(n + 1)/(2n + 1)$ which lies between 0.505 and 0.515.

6. First, pair the natural numbers with pairs of natural numbers (p, q) with $q \neq 0$, by listing the pairs thus:

$$(0, 1), (0, 2), (1, 1), (0, 3), (1, 2), (2, 1), (0, 4), (1, 3), \ldots$$

(where all those pairs whose sum of co-ordinates is n come before all those whose sum of co-ordinates is $n + 1$, and those with a fixed sum are listed in the order of their first co-ordinate).

Then we pair (p, q) with the rational p/q, leaving out any rational which has appeared before. Thus, the list of rationals begins:

$$\frac{0}{1}, \frac{1}{1}, \frac{1}{2}, \frac{2}{1}, \frac{1}{3}, \ldots$$

7. Bend the open interval (0, 1) into a semicircle and pair it with the real line as shown. (A formula could be derived from this diagram, if desired.)

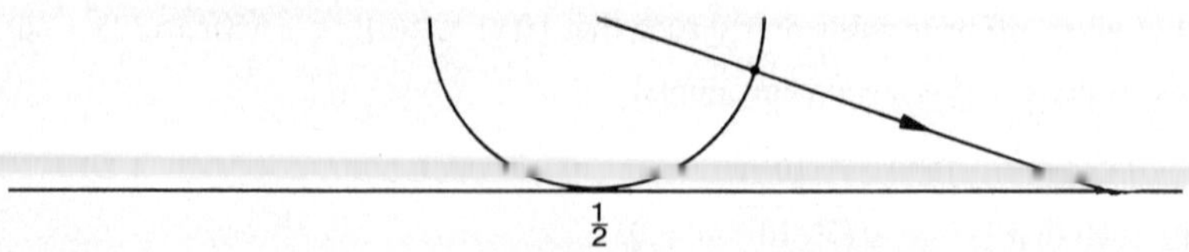

8. A proof of this result can be found in most textbooks on probability and discrete mathematics, for example, K. Kalmanson, *An Introduction to Discrete Mathematics and its Applications*.[1]

Chapter 12

1. (a) A function $f: A \to B$ is onto if for all $b \in B$ there exists $a \in A$ such that $b = f(a)$.
   (b) No. There does not exist $x \in \mathbf{R}$ such that $\sin x = 2$.
   (c) Yes. $\sin(-\pi/2) = -1$ and $\sin(\pi/2) = 1$ and $\sin x$ is continuous, so $\sin x$ takes on all the values between -1 and 1.
   (d) The range equals the codomain.

2. (a) Yes. Every $a \in \mathbf{R}$ can be written as $f(0, -a)$.
   (b) No. (1, 1) is not in the range of f, since if $f(t) = (1, 1)$ then $(1 + t, t + t^2) = (1, 1)$.

   Equating the first co-ordinate gives $t = 0$, but $f(0) = (1, 0) \neq (1, 1)$.

3. (b) No.
   (c) Yes.

4. (a) Yes. If $f(x, y) = f(x', y')$ then $(x + y, x, y) = (x' + y', x', y')$.

   So $x = x'$ and $y = y'$.

   So $(x, y) = (x', y')$.
   (b) No. $f(0, 1) = f(1, 0)$.

1. K. Kalmanson, *An Introduction to Discrete Mathematics and its Applications* (1987), Addison Wesley, Reading, Mass., pp. 124-5.

7. (a) No. It is not onto.
   (b) Yes.

9. (a) For each $a \in A$, there are m choices for f(a), namely the m elements of B. So the number of different functions is $m \times m \times \ldots \times m$ (n times), which is m^n.
   (b) Let $A = \{a_1, a_2, \ldots, a_n\}$. There are m choices for $f(a_1)$; for $f(a_2)$ there are $m - 1$ choices, since $f(a_2)$ cannot equal $f(a_1)$; and so on.

10. Yes. Any $a \in \mathbf{R}$ is F(f), where f is the constant function $f(x) = a$.

Chapter 13

1. $7\frac{1}{2}$ years.

2. Just under 35 years.

3. No. It will be about $1.0202P_0$ because of interest on interest.

6. (a)

$$f(i) = f(0) + \frac{B(A^i - 1)}{A^i(A - 1)}.$$

   (b) 15 years.
   (c) 0.35 grams/litre (this is the eventual concentration each week *before* the water is changed).

7. 25 minutes.

8. 23.8 million.

9. (a) 1 per cent.
   (b) 19 years.
   (c) The pollutant would quite probably not be well mixed; floods would wash out the lake; complaints about dead fish would occur very soon.

10. $r/\sqrt{2}$.

12. Leonhard Euler's brilliant but easily understood solution to this problem can be read in Scientific American, July 1953; it is reprinted in *Mathematics: An Introduction to its Spirit and Use.*[2]

15. Some discussion of this can be found in Thompson's *On Growth and Form.*[3] Problems of heat dissipation should be considered as well.

Chapter 14

1. $n/(n + 1)$.

2. $\frac{1}{6}n(n + 1)(2n + 1)$.

2. M. Kline (ed.), *Mathematics: An Introduction to its Spirit and Use*, Freeman, San Francisco, 1979.
3. D'A. W. Thompson (1961), *On Growth and Form* (abridged ed.), Cambridge University Press, Cambridge, pp. 20, 29.

3. $\frac{1}{4}n^2(n + 1)^2$.

4. 3, 4 and any integer greater than or equal to 6.

5. The number of segments with differently labelled ends (AB or BA) is odd; this can be proved by mathematical induction.

Appendix

1.
```
sum ← 2
i ← 2
WHILE i ≤ 2n DO i ← i+2, sum ← sum+i
STOP
```

Then for all positive integers n, the program stops with $2 + 4 + 6 + \ldots + 2n$ in sum. The proof is similar to the one in the text.

3. If N is 0 the program never stops.

Index

addition of vectors, 86–7
algorithms, 108, 131
"all" quantifier, 39
"all" statements, 3, 12ff, 39
    "All As are Bs" form, 13
    disproof of, 48
    equivalent ways of expressing, 17
    proof of, 13ff
Aristotle, 143–4, 146
arithmetic progression, 4–5, 83
assertion, 7, 28
assignment statement, 150
associative law, 9, 87, 98
axioms, 3, 87, 91, 143–5
axis of symmetry, 23, 37, 46

bijection, bijective, 128–9
bounded set, 73

calculator, 2, 4, 9, 108
calculus, 19–20, 24, 31, 35, 38, 44, 56, 71, 74, 85, 100, 104
Cantor, 120
Cauchy, 43, 146
certainty and proof, 1, 140, 143
Chinese, 149
circular argument, 156
closure under addition, 92–3
closure under scalar multiplication, 92–3
codomain, 124, 129
combinatorial explosion, 118
commutative law, 3, 87
complement of a set, 62, 117
compound interest, 131–2
computers, 2–3, 108, 118, 147, 150–3
conjectures, 9, 82, 140–2
constructive existence proof, 32, 34
continuity, 104–108
continuous function, 34, 42, 104
contradiction, proof of, 50ff, 126
contradictory of a statement, 48
contrapositive, 49, 127
convergence, 100ff
convergent series, 108–109
converse, 17, 25
convex set, 73
correspondence, one-to-one, 116–17, 128
counterexample, 48ff, 95, 126, 128
counting, 26, 116ff

De Morgan Laws, 73
deductive logic, 140
dependent, linearly, 35, 38, 44, 54–5
direct proof, 49, 57
disproof by counterexample, 48, 95
disproof of "All" statement, 48
distributive law, 87
divergent series, 108–10

empty set, 67
equality,
    functions, of, 124
    logic of, 91
    sets of, 61, 63
Equivalent statements, 20, 25, 49–50, 53, 63
Euclid, 40, 143–4, 146
Euclidean geometry, 143ff
Euler, 26–7, 141
even numbers, 20, 26, 39, 50, 64
existential generalisation, 32
existential quantifier, 39
    symbol for, 42
exponential decay, 133

factorial, 5
fallacious argument, 7–8
falling bodies, 134–6
feedback, 134
Fermat's Infinite Descent, 80
Fermat's Last Theorem, 142
Fermat's Small Theorem, 85
Fibonacci sequence, 78
formal logic, 146
four-colour theorem, 147
Frege, 146
Function, 123ff
    bijective, 128–9
    codomain of a, 124, 129
    continuous, 34, 42, 104
    image of a, 125
    injective, 126–7
    limit of a, 100–103, 114
    "many-to-one", 126
    one-to-one, 126–7
    onto, 125–6
    range of a, 125

surjective, 125–6
vector, as a, 99
Fundamental theorem of algebra, 42

Galileo, 119, 124
Gauss, 42
generalisation,
existential, 32
universal, 12, 21
geometrical progression, 37, 84
geometry, proofs in, 1, 22–3, 59, 143–4
global and local structure, 132–3
Gödel, 145–6
Golden ratio, 10

harmonic series, 108
Heaviside, 149
Hilbert, 145

ideal, 65
"if and only if" statement, 25
equivalent ways of expressing, 25
"if ... then ... and conversely" form, 30
"necessary and sufficient" form, 25
proof of, 26–9
"If" statement, 13, 17–18
equivalent ways of expressing, 17
proof of, 13ff
"iff", 25
Image
element, of an, 123
function, of a, 125
implication, 17
independent, linearly, 55
induction, 82, 140–2
mathematical, 76ff
infinite series, 37–8, 108
infinity, 101–102, 119–20
primes, of, 40
injective function, 126–7
insight, 143, 147
intersection of sets, 62
invariant, loop, 152
irrational numbers, 46, 48, 51–2
irrationality of $\sqrt{2}$, e and π, 51–2

Leibniz, 62
limit, 100ff
function at a point, of a, 114
function at infinity, of a, 100–103
sequence, of a, 103–104
linear algebra, 18, 24, 31, 34, 38, 43, 54, 68, 86ff
linear combinations, 18–19, 24, 34, 43, 54, 68
linear equations, 18, 24
linearly dependent, 35, 38, 44, 54
linearly independent, 55
local and global structure, 132–3
logic, 1, 3, 12, 17, 43–5, 131, 143
deductive, 140
non-deductive, 140–1, 144
symbolic, 146–7
logically equivalent, 20, 25, 49–50, 53, 63

"many-to-one" function, 126
mathematical induction, 76
mathematics, 1, 7, 15, 145
matrices as vectors, 87
modelling, mathematical, 131–6
multiple, 32
multiple quantifiers, 39–45, 100ff
and "not", 53

necessary and sufficient condition, 25
necessary condition, 17
negation, 48
non-constructive existence proof, 33, 41
non-deductive logic, 140–1, 144
non-Euclidean geometry, 144–5
"not-all" statement, 48ff, 70
proof of, 48

Numbers
even, 20, 26, 39, 50, 64
irrational, 46, 48, 51–2
odd, 15ff, 50
perfect, 36
prime, 30, 37, 40, 80
rational, 25, 31, 46, 51
whole, 3, 12, 26, 32, 39–41, 76

obvious truths, 1, 3, 5–7, 143–7
odd numbers, 15ff, 50
one-to-one correspondence, 116–17
one-to-one function, 126–7
"only if" statements, 17
onto function, 125–6
Oresme, 110

pairing, 116ff, 128
particular statements, 12
Peano axioms, 145
Perfect number, 36
pi, 11
pigeonhole principle, 85, 159
Platonic solids, 25–6
polyhedra, 25–6
polynomials as vectors, 86–9
population growth, 136–7

predicate, 12
prime number, 30, 37, 40, 80
progression
  arithmetic, 4–5, 83
  geometric, 37, 84
proof, 1ff
  "all" statement, of, 13ff
  computer generated, 144
  construction of a, 1–2
  constructive existence, 32, 34
  contradiction, by, 50ff
  counterexample, by, 48ff, 95
  counting, by, 26, 116ff
  direct, 48, 57
  enumeration, by, 116
  fallacious, 7–8, 84
  from axioms, 3, 91, 143
  "if and only if" statement, 26–9
  "if" statement, of, 13ff
  mathematical induction, by, 76ff
  non-constructive existence, 33, 41
  "not-all" statement, 48ff
  reductio ad absurdum, by, 50ff, 120
  search for a, 8
  set equality, of, 63–7
  set inclusion, of, 63–4
  "some" statement, of, 32
  uniqueness, 89–90
proposition, 50, 78, 144
Pythagoras' theorem, 1, 22, 23, 144

quantifiers, 39ff
  existential, 39
  interchange of, 42–3
  multiple, 39–45
  universal, 39

range of a function, 125
rational numbers, 25, 31, 46, 51
reductio ad absurdum proof, 50ff, 120
regular polyhedron, 25
relation, 12
reversible steps, 7, 29
Rolle's theorem, 60
round-off errors, 3
Russell, B., 145–6

scalar, 24, 35, 43, 44, 87ff
scalar multiple, 24, 38, 44
scalar multiplication, 74, 86ff
search for a proof, 8
sequence, limit of a, 103–104, 112
series, 108
  convergent and divergent, 108–10
  for e^x, 38
  harmonic, 108–10
sets, 17, 61ff
  bounded, 73
  complement, 62
  convex, 73–4
  element of, 61
  intersection and union of, 62
"some" statements, 32
span, 68–70
stability, 134
statement,
  "all", 3, 12ff, 39
  assignment, 150
  existential, 80
  "if", 17ff
  "if and only if", 25ff
  "iff", 25
  "not-all", 48, 70
  "only if", 17
  particular, 12
  "some", 32
subject, 12
subsets, 62–3
  number of, 80–1, 116–17
subspace, 74, 92–5
sufficient condition, 17
sum of a series, 108
surjective function, 125–6
symbolic logic, 146–7
symmetry
  axis, about an, 23, 37, 46
  point, about a, 23

truths
  mathematics, in, 1
  obvious, 1, 3, 5–7, 143–7

union of sets, 62
uniqueness proof, 89–90
universal generalisation, 12, 21
universal quantifier, 39
  symbol for, 42
vector space, 86ff
vectors, 18–19, 34–5, 38, 43ff, 55, 68ff, 86ff
  addition of, 87
  functions as, 99
  matrices as, 87
  scalar multiplication of, 87
Venn diagrams, 48, 62

Weierstrass, 146
Whitehead, 146
whole numbers, 3, 12, 26, 32, 39–41, 76

zero vector, 44, 87–90